Landslides

WITPRESS

WIT Press publishes leading books in Science and Technology.
Visit our website for the current list of titles.
www.witpress.com

WITeLibrary

Home of the Transactions of the Wessex Institute, the WIT electronic-library provides the international scientific community with immediate and permanent access to individual papers presented at WIT conferences.
Visit the WIT eLibrary athttp://library.witpress.com

Safety & Security Engineering Series
Series Editor: S. Mambretti, *Politecnico di Milano, Italy*

Titles in Series

Flood Risk Assessment and Management
Landslides

Landslides

Edited by

S. Mambretti
Politecnico di Milano, Italy

Safety & Security Engineering Series

Series Editor: S. Mambretti, *Politecnico di Milano, Italy*

SAFETY &
SECURITY
ENGINEERING

Editor: S. Mambretti, *Politecnico di Milano, Italy*

Published by

WIT Press
Ashurst Lodge, Ashurst, Southampton, SO40 7AA, UK
Tel: 44 (0) 238 029 3223; Fax: 44 (0) 238 029 2853
E-Mail: witpress@witpress.com
http://www.witpress.com

For USA, Canada and Mexico

WIT Press
25 Bridge Street, Billerica, MA 01821, USA
Tel: 978 667 5841; Fax: 978 667 7582
E-Mail: infousa@witpress.com
http://www.witpress.com

British Library Cataloguing-in-Publication Data

A Catalogue record for this book is available from the British Library

ISBN: 978-1-84564-650-9
eISBN: 978-1-84564-651-6
ISSN (print): 2047-7686
ISSN (online): 2047-7694

Library of Congress Catalog Card Number: 2011937216

The texts of the papers in this volume were set individually by the authors or under their supervision.

No responsibility is assumed by the Publisher, the Editors and Authors for any injury and/or damage to persons or property as a matter of products liability, negligence or otherwise, or from any use or operation of any methods, products, instructions or ideas contained in the material herein. The Publisher does not necessarily endorse the ideas held, or views expressed by the Editors or Authors of the material contained in its publications.

© WIT Press 2012

Printed in Great Britain by Polestar Wheatons.

All rights reserved. No part of this publication may be reproduced, stored in a retrieval system, or transmitted in any form or by any means, electronic, mechanical, photocopying, recording, or otherwise, without the prior written permission of the Publisher.

Preface

This volume is the second of the Series the Publisher decided to launch to cover different aspects related to Safety and Security Engineering.

The volume is devoted to landslides; this geological phenomenon includes a wide range of ground movement, such as rock falls, deep failure of slopes and shallow debris flows. Although the action of gravity is the primary driving force for a landslide to occur, there are other contributing factors affecting the original slope stability. Typically, pre-conditional factors build up specific sub-surface conditions that make the area/slope prone to failure, whereas the actual landslide often requires a trigger before being released.

The phenomenon is of great importance as it represents some of the most destructive forces on earth. Therefore it is imperative to have a good understanding as to what causes them and how people can either help prevent them from occurring or simply avoid them when they do occur. Early predictions and warnings are essential for the reduction of property damage and loss of life. Sustainable land management and development is an essential key to reducing the negative impacts felt by landslides.

The mentioned topics are covered in the present volume, and in particular:

- Risk assessment
- Early Warning
- Modelling
- Planning
- Mitigation
- Case studies

The volume contains selected papers presented at the Conferences organized by the Wessex Institute of Technology, which have been revised by the Authors, in order to be up-to-date and integrated in the book.

The Editor is very grateful to the Authors for their excellent contribution, and to the WIT Press for the availability in continuing to publish this Series in these very important subjects. The quality of the material makes the volume a very good overview of the present state of the art in landslides and a most valuable and up-to-date tool for professionals, scientists, and managers to appreciate the state-of-the-art in this important field of knowledge.

The Editor
Milano, Italy
2012

Contents

Multi-hazard analysis in natural risk assessments
R. Bell & T. Glade 1

From national landslide database to national hazard assessment
M. Jemec & M. Komac 11

On the definition of rainfall thresholds for diffuse landslides
L. Longoni, M. Papini, D. Arosio & L. Zanzi 27

Computer analysis of slope failure and landslide processes caused by water
I. Sarafis & J. Zezulak 45

The integration between planning instruments and evaluation tools in the
management of landslide risk
M. Magoni 55

Finite element analysis of the stability of artificial slopes
reinforced by roots
F. Gentile, G. Elia & R. Elia 63

Experience with treatment of road structure landslides by
innovative methods of deep drainage
O. Mrvík & S. Bomont 79

Strategic program for landslide disaster risk reduction:
a lesson learned from Central Java, Indonesia
D. Karnawati, T. F. Fathani, B. Andayani, P. W. Burton & I. Sudarno 91

Erosion of forestry land: cause and rehabilitation
T. Ogawa, Y. Yamada, H. Gotoh & M. Takezawa 103

Landslide in a catchment area of a torrent and the consequences for
the technical mitigation concept
F. J. Riedl 115

Slope instability along some sectors of the road to La Bufadora
J. Soares, C. García, L. Mendoza, E. Inzunza, F. Jáuregui & J. Obregón...... 125

Multi-hazard analysis in natural risk assessments

R. Bell & T. Glade
Department of Geography, University of Bonn, Germany

Abstract

Analysis of natural risks in mountainous regions includes several typical natural processes such as snow avalanches, floods, earthquakes, and different types of landslides. Separate investigations of single processes only might lead to a misjudgement of the general natural risks for these areas. To avoid this trap, natural risk assessments should not focus on a singular process but on multiple processes. Within this study a general methodology is developed to analyse natural risk for multiple processes. The method is applied in Bíldudalur, NW-Iceland. In particular snow avalanches, rock falls and debris flows pose a hazard to the village of 300 inhabitants. The natural risk calculation is a function based on the input parameters hazard, vulnerability, probability of the spatial impact, probability of the temporal impact, probability of the seasonal occurrence and damage potential. First, the risk posed by each process is calculated. Results are presented as individual risk and object risk to life, and as economic risk for each process. Finally, single process risk maps are combined into multi-hazard risk maps. In the study area the highest risks throughout all of the analyses (individual risk to life, object risk to life and economic risk) are caused by debris flows, followed by snow avalanches and rock falls. It is demonstrated that risk varies heavily depending on the process considered. The total risk to life caused by snow avalanches, debris flows, rock falls and multi-hazards is 0.19, 0.63, 0.009 and 0.83 deaths per year, respectively. Multi-hazard approaches are not only valuable to get an overview on the overall risk but have also a high significance for planning effective countermeasures. It can be concluded that the newly developed method is applicable to other natural processes as well as to further catchments in Iceland as well as in other countries with different environmental settings.

Keywords: natural hazards, risk assessment, snow avalanches, debris flows, rock falls, Iceland.

1 Introduction

"Society in general and individuals within it all face various risks. These cannot be eliminated, only reduced by applying additional resources. Furthermore, the reduction of risks from one hazard may increase risks from other hazards, and thus not be beneficial overall" [1].

The example of Gondo (Switzerland) demonstrates this very well. To mitigate rock fall hazard a combined wall/fence structure was build to collect falling rocks and prevent to threat the community of Gondo any longer. Unfortunately, on 14 October 2000 extreme precipitation triggered a debris flow, which, first, was caught by the rockfall mitigation structure, but later, the pressure caused by the debris flow material exceeded the withstand-power of the mitigation structure which then failed and released all material at once, taking with it the material of the structure and then moved into the village and destroyed several houses and caused 13 deaths. Without the rock fall mitigation structure the event potentially could have been less severe. This case shows that countermeasures against one hazardous process (here, rockfall) can increase the threat of another process (here, debris flow) [2]. Demands resulting from such a disaster are that multi-hazard risk assessments should be always carried out whenever possible and should include calculation under natural conditions as well as considering counter-measures.

Following these demands, a general methodology is developed to analyse natural risk for mutli hazards within this study. The method is applied in Bíldudalur, NW-Iceland, where in particular snow avalanches, debris flows and rock falls pose a threat to the village of 300 inhabitants.

2 Risk assessment

Usually, when natural disasters occur both environmental and human systems are involved. Natural events do not pose a threat to society or a community if the affected area is not used by people. Thus, holsitic concepts are necessary to analyse the complex interactions between these two systems and to find the "best" solutions for endangered areas adopted to local needs.

Hollenstein developed such a holistic concept to natural risks [3]. The entire risk assessment consists of three equal parts: risk analysis, risk evaluation and risk management. The main focus of each part is demonstrated by the questions given in Figure 1. For more details refer to [3, 9].

For specific risks, such concepts are provided by numerous authors, e.g. for landslide risk by [4–8].

Within this study, risk analysis alone is considered.

3 Study area "Bíldudalur"

The village Bíldudalur is located in the Westfjords in NW-Iceland (Figure 2a). Especially snow avalanches, debris flows, rock falls and slush flows pose a threat to the community (see [9] or [13] for more details). Within this study, only the former three could be investigated.

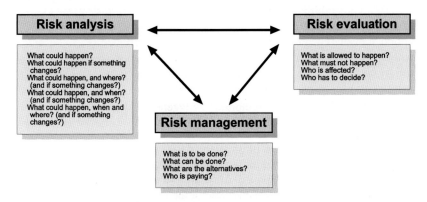

Figure 1: The holistic concept of risk assessment (based on [3, 10–12]).

The Westfjords are characterized by a fjord landscape with typical u-shaped valleys. Extensive plateaus can be found on the top of the mountains, which rises up to 460m a.s.l. above Bíldudalur. The mountainside is dissected into two large gullies and several smaller ones in between (Figure 2b), followed by respective debris cones. The mild and maritime climate is characterized by cool summers and mild winters. Mean annual air temperature is 3°C and annual precipitation ammounts to approximately 1250mm. The lithology consists of various basaltic layers, which are nearly horizontal bedded. Periglacial, gravitational and fluvial processes are dominating the study area (for more details refer to [9] or [13]).

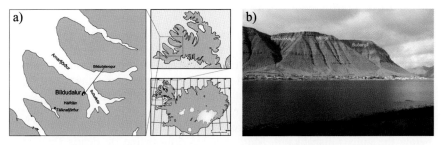

Figure 2: Study area Bíldudalur - a) location, b) photography from opposite Fjord border, view towards the north.

4 Methodology

Within this study, a new raster based approach on a regional scale was developed based on recent approaches to risk analysis [6,11,14–16]. The approach consists of the following steps: scope definition, hazard identification, hazard analysis, consequence analysis and risk calculation.

The calculation of the natural risk follows a function of the input parameters hazard (H), vulnerability (of people (V_{pe}), property (V_p), infrastructure (V_{str}) and powerline (V_{po})), probability of the spatial impact (P_s), probability of the temporal impact (P_t), probability of the seasonal occurrence (P_{so}) and damage potential (number of people (E_{pe} or E_{ipe}), economic value (E_p)). First, the risk posed by each process is calculated. Results are presented as individual risk and object risk to life, and as economic risk for each process. Finally, single process risk maps are combined to multi-hazard risk maps. A detailed description of the methodology in general and for debris flows and rock falls in particular is given in [9].

For snow avalanches more information must be provided. Snow avalanche risk analysis is based on a preliminary snow avalanche hazard map created by Siegfried Sauermoser, who applied Austrian guidelines to delineate hazardous areas. Two different hazard zones resulted: a red hazard zone and a yellow zone. The border of the red hazard zone is defined as a snow avalanche with a return period of 150 years exceeding an impact pressure of 10 kN/m² (until recently the threshold was 25 kN/m²) or as a snow avalanche with a return period of 10 years on the average. The border of the yellow hazard zone is defined as a 150 year event exceeding an impact pressure of 1 kN/m² (Sauermoser 2002, personal communication).

Figure 3 shows the methodological concept of multi-hazard analysis including the respective formulas to calculate the individual risk to life, the object risk to life and the economic risk.

5 Results

5.1 Hazard identification and analysis

Snow avalanches, debris flows and rock falls pose threats to people, properties and infrastructure along the whole length of the village. The highest snow avalanche and debris flow hazards exist below the two large gullies (refer to figure 2b). However, also the smaller catchments in between these two large gullies are active. Recently more but smaller debris flow events were triggered from these small gullies. Regarding rock falls field investigations show that the most north-eastern part is most active, while the largest boulder could be found below the gully Gilsbakkagil. More detailed information on debris flows and rock falls is given in [9].

5.2 Consequence analysis

The spatial pattern of the economic value of the elements at risk is given in Figure 4. As detailed data is confidential, following four classes were defined: very low (0-36€/m²: 16 buildings and the power line), low (>36-480€/m²: roads, infrastructure and 45 buildings), medium (>480-960€/m²: 72 buildings), high

(>960-1440€/m²: 26 buildings) and very high (>1440€/m²: 13 buildings). The spatial pattern of residents and employees is given in [9].

The vulnerability values are determined based on the process and its magnitude (debris flow and rock fall) or hazard (snow avalanches). Values used are presented in table 1.

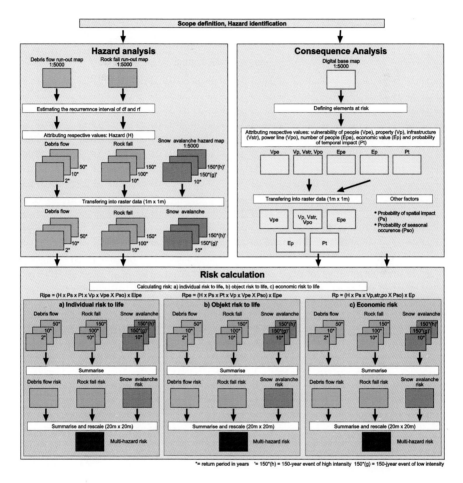

Figure 3: Methodological concept of multi-hazard analysis.

Applied values for the probability of spatial impact are shown in table 2. As the hazards map show, even large debris flows or snow avalanches would not affect the whole settlement. Therefore, low values were estimated.

Regarding the probability of temporal impact (i.e. of the building being occupied given an event), for residential houses 18h a day was chosen, whereas for companies and the school a common value is 9-10h a day.

6 Safety & Security Engineering

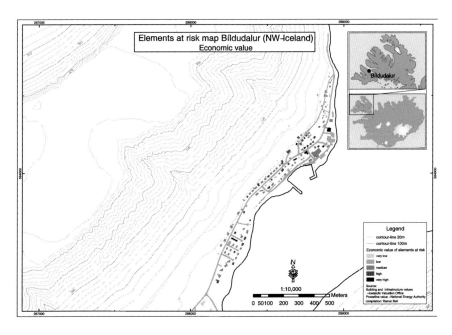

Figure 4: Elements at risk map – economic value.

Table 1: Vulnerability values used within this study (Note: V_{po} = vulnerability of the power line, V_{str} = vulnerability of roads and infrastructures, V_{p} = vulnerability of properties, V_{pe} = vulnerability of people and V_{pep} = vulnerability of people in buildings, high(1) = 10 year event of high hazard class, high(2) = 150 year event of the high hazard class).

Magnitude	low					medium					high				
Process	V_{po}	V_{str}	V_p	V_{pe}	V_{pep}	V_{po}	V_{str}	V_p	V_{pe}	V_{pep}	V_{po}	V_{str}	V_p	V_{pe}	V_{pep}
Debris flow	1.0	0.2	0.1	0.2	0.02	1.0	0.4	0.2	0.3	0.06	1.0	0.6	0.5	0.5	0.25
Rock fall	1.0	0.1	0.1	0.2	0.02	1.0	0.2	0.3	0.4	0.12	1.0	0.4	0.5	0.5	0.25
Hazard	low					high(1)					high(2)				
Process	V_{po}	V_{str}	V_p	V_{pe}	V_{pep}	V_{po}	V_{str}	V_p	V_{pe}	V_{pep}	V_{po}	V_{str}	V_p	V_{pe}	V_{pep}
Snow avalanche	1.0	0.3	0.3	0.5	0.15	1.0	0.1	0.3	0.1	0.03	1.0	0.8	1.0	1.0	1.0

Since snow avalanches only occur in winter times the probability of seasonal occurrence is set to 0.5. For debris flows and rock falls the factor 1 is chosen due to the fact that both processes can occur during the whole year, as historical records demonstrate.

The final individual risks to life, object risks to life and economic risks due to snow avalanches, debris flows, rock falls and multi-hazards are summarised in Table 3. The total risk to life caused by snow avalanches, debris flows, rock fall and multi-hazards is 0.19, 0.63, 0.009 and 0.83 deaths per year, respectively. Figure 5 presents the economic risks posed by multi-hazards.

Table 2: Probability of spatial impact of each process dependant on its magnitude or hazard (Note: within snow avalanches, high(1) refers to the criterion of the event with a return period of 10 years and high(2) is related to the 150 year event. As stated in chapter 4 low hazard refers also to the 150 year event but with a lower impact pressure. Since the 10 year event refering to high(1) hazard is supposed to be smaller than the 150 year event refering to low hazard, the lowest value is chosen for high(1) hazard.).

Magnitude / Process	low	medium	high
Debris flow	0.1	0.2	0.3
Rock fall	0.01	0.01	0.02

Hazard / Process	low	high(1)	high(2)
Snow avalanche	0.3	0.2	0.5

Table 3: Individual risk to life, object risk to life and economic risk in Bíldudalur.

Risk type	Unit	risk value		% per risk class			
		min	max	very low	low	medium	high
individual risk to life				$<0.3*10^{-4}$	$0.3 - <1.0*10^{-4}$	$1.0 - <3.0*10^{-4}$	$>3.0*10^{-4}$
snow avalanche	r/a	5.6×10^{-5}	1.6×10^{-3}	0.00	33.53	26.47	40.00
debris flow	r/a	5.7×10^{-4}	2.8×10^{-3}	0.00	0.00	0.00	100.00
rock fall	r/a	1.1×10^{-5}	5.6×10^{-5}	7.80	92.20	0.00	0.00
multi-hazard	r/a	5.7×10^{-5}	4.4×10^{-3}	0.00	0.00	7.08	83.63
object risk to life				$<0.3*10^{-4}$	$0.3 - <1.0*10^{-4}$	$1.0 - <3.0*10^{-4}$	$>3.0*10^{-4}$
snow avalanche	r/a	6.3×10^{-5}	2.9×10^{-2}	0.00	14.72	21.18	64.12
debris flow	r/a	6.3×10^{-4}	7.8×10^{-2}	0.00	0.00	0.00	100.00
rock fall	r/a	2.1×10^{-5}	1.6×10^{-3}	4.26	26.95	57.45	11.35
multi-hazard	r/a	6.3×10^{-5}	8.2×10^{-2}	0.00	4.42	3.10	92.48
economic risk				<3.6	$3.6 - <9$	$9 - <18$	$>=18$
snow avalanche	€/m²/a	0.024	9.84	4.26	26.95	57.45	11.35
debris flow	€/m²/a	0.24	26.52	42.09	46.28	9.77	1.86
rock fall	€/m²/a	0.0036	0.22	100.00	0.00	0.00	0.00
multi-hazard	€/m²/a	0.036	33.84	50.67	24.38	21.69	3.26

6 Discussion

Final results show that snow avalanches, debris flows and rock falls pose partly serious threats to the community of Bíldudalur.

The highest risks by far throughout all of the analyses (individual risk to life, object risk to life, economic risk) are caused by debris flows, followed by snow avalanches and rock falls. The low return periods of 2, 10 and 50 years of the debris flows lead mainly to the high debris flow risks. Further investigations are necessary to improve the reliability of the return periods (e.g. sediment supply rates must be defined in detail, see [9] and [17] for more details). The calculated risk in relation to snow avalanches seems to be more reasonable, since higher recurrence intervals were applied in the hazard map and in the risk calculations. Rock falls are very local phenomena. Thus, the probability of spatial impact is very low, causing relatively low values of rock fall risks. However, this does not mean that rock falls might not cause economic damage or fatalities in the study area.

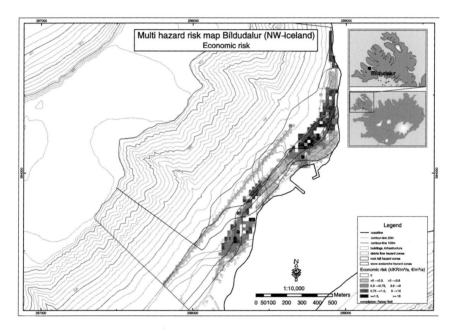

Figure 5: Multi-hazard risk map – economic risk.

The multi-hazard risks give an indication of the overall risk posed to the community. Multi-hazard approaches are not only valuable to get an overview on the overall risk but have also a high significance for planning effective countermeasures. To avoid the trap of reducing risks from one hazard, but increasing risks from other hazards, as shown by the example of Gondo, multi-hazard analyses should be more often applied within natural risk assessments.
It can be concluded that the newly developed method is applicable to further processes as well as to further catchments in Iceland, but also to other countries with different environmental settings.

Acknowledgements

We are very grateful to E. Jensen and the staff of the Icelandic Meteorological Office for financial support and the provision of data. Without their assisstance, this study could never be carried out in the presented form. Finally, we would like to thank S. Sauermoser for the provision of the preliminary snow avalanche hazard map.

References

[1] Finlay, P.J. & Fell, R., Landslides: Risk perception and acceptance. *Canadian Geotechnical Journal*, **34(2)**, pp. 169-188, 1997.

[2] Federal Institute for Water and Geology (BWG), Medienorientierung zur Ereignisanalyse Hochwasser 2000 http://www.bwg.admin.ch/aktuell/presse/2002/d/fbhw2000.htm (09-06-04), 2002.

[3] Hollenstein, K., *Analyse, Bewertung und Management von Naturrisiken*, vdf Hochschulverlag AG, ETH Zürich: Zürich, pp. 220, 1997.

[4] Fell, R., Landslide Risk Management Concepts and Guidelines - Australian Geomechanics Society Sub-Committee On Landslide Risk Management. Landslides, Cardiff, UK, ed. *International Union of Geological Sciences*, International Union of Geological Sciences, pp. 51-93, 2000.

[5] Einstein, H.H., Landslide risk - systematic approaches to assessment and management. eds. *Cruden, D.M. & Fell, R.*, A.A. Balkema: Rotterdam, pp. 25-50, 1997.

[6] Fell, R. & Hartford, D., Landslide risk management. eds. *Cruden, D.M. & Fell, R.*, A.A. Balkema: Rotterdam, pp. 51-109, 1997.

[7] Leroi, E., Landslide hazard - Risk maps at different scales: Objectives, tools and development. Landslides - Glissements de Terrain, 7th. International Symposium on Landslides, Trondheim, Norway, ed. *Senneset, K.*, Balkema, pp. 35-51, 1996.

[8] Einstein, H.H., Special lecture: Landslide risk assessment procedure. Proceedings of the 5th International Symposium on Landslides, 10-15 July 1988, Lausanne, Switzerland, ed. *Bonnard, C.*, A.A. Balkema, pp. 1075-1090, 1988.

[9] Bell, R. & Glade, T., Quantitative risk analysis for landslides - Examples from Bíldudalur, NW-Iceland. *Natural Hazard and Earth System Science*, **4**, pp. 1-15, 2004.

[10] Glade, T., Landslide hazard assessment and historical landslide data - an inseparable couple? *Advances in Natural and Technological Hazards Research*. eds. *Glade, T., Frances, F. & Albini, P.*, Kluwer Academic Publishers: Dordrecht, pp. 153-168, 2001.

[11] Heinimann, H.R., *Risikoanalyse bei gravitativen Naturgefahren - Methode*: Bern, pp. 115, 1999.

[12] Kienholz, H., Naturgefahren - Naturrisiken im Gebirge. Forum für Wissen, Naturgefahren, Birmensdorf, Switzerland, eds. *Eidg. Forschungsanstalt fuer Wald, S. & Landschaft, u.*, pp. 7 - 21, 1993.

[13] Glade, T. & Jensen, E.H., Landslide hazard assessments for Bolungarvík and Vesturbyggð, NW-Iceland. Icelandic Metereological Office: Reykjavik, 2004.

[14] Glade, T., von Davertzhofen, U. & Dikau, R., GIS-based landslide risk analysis in Rheinhessen, Germany. *Natural Hazards*, subm.

[15] Fell, R., Landslide risk assessment and acceptable risk. *Canadian Geotechnical Journal*, **31(2)**, pp. 261-272, 1994.

[16] Morgan, G.C., Rawlings, G.E. & Sobkowicz, J.C., Evaluating total risk to communities from large debris flows. Geotechnique and Natural Hazards, 6 - 9 May 1992, Vancouver, Canada, pp. 225-236, 1992.

[17] Glade, T., Linking natural hazard and risk analysis with geomorphology assessments. *Geomorphology*, 2004, in press.

From national landslide database to national hazard assessment

M. Jemec & M. Komac
Geological Survey of Slovenia, Slovenia

Abstract

The territory of Slovenia is, geologically speaking, very diverse and mainly composed of sediments or sedimentary rocks. Slope mass movements occur in almost all parts of the country. In the Alpine carbonate areas of the northern part of Slovenia rockfalls, rock slides and even debris flows can be triggered. In the mountainous regions of central Slovenia composed from different clastic rocks, large soil landslides are quite common, and in the young soil sediments of the eastern part of Slovenia there is a large density of small soil landslides. The damage caused by slope mass movements is high, but still no common strategy and regulations to tackle this unwanted event, especially from the aspect of prevention, have been developed. One of the first steps towards an effective strategy of combating landslides and other slope mass movements is a central landslide database, where ideally all known landslide occurrences would be reported and described in as much detail as possible. At the end of the project of compiling the National Landslide Database (May 2005) there were more than 6,600 registered landslides, of which almost half occurred at a known location and were accompanied with the main characteristic descriptions. Based on the landslide database described, a Landslide Susceptibility Map of Slovenia at a scale 1:250,000 was completed. Of 3,257 landslides with a known location, a random but representative 65% were selected and used for the statistical analysis of landslide occurrence, the rest of the landslide population (35%) being used for model validation. The most suitable susceptibility model was used for the anthroposphere exposure assessment due to potential landslide processes on a national scale. These analyses included a population census, building locations, land use, road type and railways. The results have shown that almost 19% of the population lives in one quarter of the area of Slovenia that is highly exposed to landslide occurrence. This is clearly an indication that better spatial and urban planning is needed on a national scale.

Keywords: landslide, slope mass movements, hazard, database, Slovenia.

1 Introduction

Based on research conducted in the early nineties, [1] estimated that there could be between 7,000 and 10,000 active landslides in Slovenia (or 0.3-0.5/km2). To put it more illustratively, there is one landslide per 1.46 square kilometres in Slovenia, excluding flat terrain. In the years 1994 to 2006, the damage caused by landslides (and avalanches) amounted to €94.2 million [2], excluding remediation costs. In the past decade global climate changes probably caused several extremely large landslide events on the territory of Slovenia, which had not been observed before. The latter represents an additional, much higher burden on the state and municipal budgets. In rare, but extreme circumstances, the landslides resulted in human casualties.

The newly arisen conditions due to climate changes demand a more strategic approach in tackling the problems related to slope mass movements. This is especially important in mountainous areas which occupy roughly one third of Slovenia. Worldwide there are numerous landslide databases, all of them with a common prerogative – landslide databases are constructed to study the evolution of landscapes, and are mandatory to ascertain landslide susceptibility, hazard and risk. In Europe, landslide databases can be found in Italy [3] and France [4], to name just a few countries. In Asia and Oceania similar projects are under way in Japan [5], Taiwan [6], Hong Kong [7], and Australia [8], while in America such projects exist in USA [9] and Canada [10]. The basis for the worldwide landslide database has been set by UNESCO [11–13]. Based on the 2006 Tokyo Action Plan, the International Programme on Landslides (IPL) Global Promotion Committee was established by the International Consortium on Landslide (ICL) members and ICL supporting organizations [14]. One of the Consortium's main goals is to maintain a database of the world's landslides [15].

2 National landslide database in Slovenia

With an awareness that Slovenia is highly exposed to landslides or rather several types of slope mass movements, several Slovenian Ministries expressed the will to finance the construction of the National Landslide DataBase [16, 17] which includes events of landslides, rockfalls and debris flows. In the following text a landslide database is referred to; the word landslide in the text should be considered as slope mass movements in general. Some of the pioneering work was done in the 1990s with pilot projects [1]. Nowadays, an up-to-date landslide database is vital for the activity of the Ministry of the Environment and Spatial Planning for gathering data on imminent danger (geohazard maps) and coping with the issues of prevention and remediation due to slope mass movement events. Usually there are huge costs related to the remediation of the consequences due to slope mass movement occurrence, which are partly compensated by the state. Additionally, the Administration of the Republic of Slovenia for Civil Protection and Disaster Relief deals with the disaster impacts on the population and their property.

The project goals were: (a) to establish an up-to-date central landslide database, (b) the construction of an information system that would allow data management by different users via the internet application. The database would represent (c) the basis for spatial analysis of slope mass movement distribution, and (d) the slope mass movement data could be distributed very fast to different users in accordance with their privileges/rights. Furthermore, the database would serve as a foundation for the modelling and production of geohazard and georisk maps of different scales (e).

The existing slope mass movement data that were acquired from different sources in different formats were analysed and merged into the centralised database with the duplicates removed. The quality is questionable to a certain degree, since the separate databases were rarely maintained. The dominating problems were different database attributes and missing or multiplied data [18]. At the end of the project there were 6,602 slope mass movements in the database, while 3,257 of them were geolocated (Fig. 1).

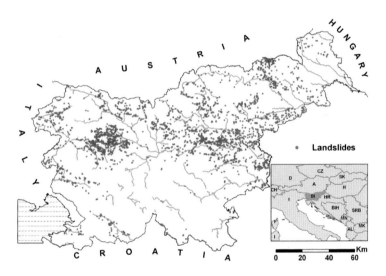

Figure 1: Known landslide distribution in Slovenia.

The Landslide DataBase is made up of the following types of data: a) Basic data (Code, Name, Location, Date of occurrence), b) Register of spatial data (Municipality, Settlement), c) Coordinates, d) Landslide conditions (Status, Speed, Dimensions, Geology), e) Remediation of Landslide, f) Costs of remediation, g) Priority, h) Documentation, i) Activities on Landslide, j) Landslide occurrence consequences (Damaged and threatened objects, Roads, Buildings, Public Infrastructure, Land).

Three different modules, the Authorization module (Managing users and their rights [user name and password], Access to application, Controlling digital certificates), the Attribute module (Landslide registration, Landslide data modification, Managing landslide events, Search and query, Managing attribute data), and the Spatial module (Viewing and checking different graphical layers, Zoom, Identification, Measuring distance, Selecting graphical objects and their transfer to the attribute module; Fig. 2) constitute the Landslide Information System (LIS). Applications supporting the users' work consist of attribute (for tabular data) and GIS (for graphical spatial data) modules connected to one system. Both modules are based on multitier internet technology. The attribute module is created with the JSP (Java Server Pages) and runs on the Apache application server with installed Oracle Container for JAVA (OC4J) and attributes are stored in Oracle database 9i2R. The GIS WEB module is developed within the Delphi environment with the ESRI MapObjects components for GIS while Oracle Spatial is used for storing the location of slope mass movements. Other spatial data used in the system are stored on the file server (background maps, digital orthophoto, etc), or in Oracle Spatial (vector data such as land and building cadastre, infrastructure, etc). Also the Internet Map Server was implemented to support communication between users and the GIS WEB module. At the user level the thin client in Java supports the GIS functionality. The users of the system can be divided into three segments: the administrator, internal users and external users (Fig. 2). All the events on the slope mass movements are stored in the History of Events files. In this way the system enables tracking of all the events on every slope mass movement entered by users in the slope mass movement registry until now. Prior to modifying the data, the type of event, reason for the change and the change date have to be defined. Every event is managed as an independent entity and can be recalled in the same form as was entered.

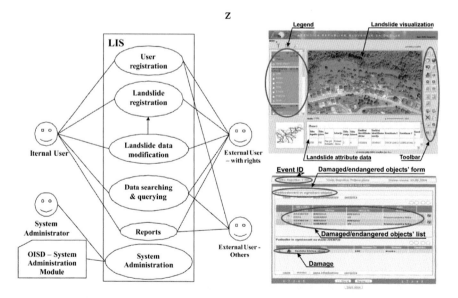

Figure 2: Flowchart of the Landslide Information System (LIS) and a part of the Attribute module (below) and Spatial module (above).

3 Methodology

3.1 National landslide susceptibility map

The constructed database enabled the spatial and temporal analyses of landslide occurrence in relation to spatio-temporal factors, and based on these results the landslide susceptibility map at a scale of 1:250,000 for the area of Slovenia was produced. All the analyses were conducted in the GIS with 25×25 m pixel resolution and the results were statistically analysed employing the χ^2 method. For the landslide susceptibility map a linear model of weighted spatial factors was used.

For the purpose of model development, the spatial factor data that had already been proven to be relevant to landslide susceptibility by many authors [19–22] were gathered. The landslide data were obtained from the National Landslide Database. Approximately 2/3 of landslides (2,176) were randomly, but representatively selected from the population for each geological engineering unit for the landslide susceptibility model training phase. The remaining 1,081 landslides were used for model evaluation. Where less than 40 landslides occurred in a specific geological engineering unit, the landslide occurrence served as an indication that assisted the expert to make the right classification decision of landslide occurrence probability for a given unit. The digital elevation model (DEM) data were obtained from the national 25 m resolution InSAR DEM 25 [23]. All the additional data on the terrain morphology (curvature, elevation, slope, aspect) were derived from the DEM. The Geological Map of Slovenia at a scale of 1:250,000 [24] served as a source for the geological engineering data [18, 25]. For land use and vegetation cover CORINE land cover data were used [26]. The surface water data were obtained from ARSO [27] and are at a scale of 1:25,000.

To understand the natural processes, the influencing spatio-temporal factors on the observed process have to be defined and their interaction addressed. The most appropriate way to understand the "back-stage" of natural processes is to analyse the factors or their approximations. The better the understanding, the better can be the prediction of future events. The groups of influencing factors on landslide occurrence in Slovenia were selected based on previous research [28, 29]. The analyses were conducted on the landslide population for all of the spatio-temporal factors for the whole of Slovenia, and additionally on the landslide population for all of the spatio-temporal factors for each of the 29 geological engineering units.

Several authors [30, 31] showed the applicability of the χ^2 (Chi-square) method for testing normally distributed discrete variables. The χ^2 method is based on a comparison of observed and expected frequencies of the phenomenon [32]. For the purpose of model development the categorical variables were transformed into numerical form on the basis of relative landslide occurrence probability of phenomenon occurrence calculated from the χ^2 values for a specific class of variable. In short, they were ordered, but one has to consider that such an ordinal scale does not comply strictly with the law of continuity.

Where obvious discrepancies of class order occurred, an expert decision was made to correct the error. Before the inclusion of relevant factors in the model development, the values of each factor were normalised. This was a necessary step to equalise the different class numbers in factors, the goal being that the weights in the models represented the real influence of a given factor. Normalisation was done using the eqn (1):

$$NV = \frac{5*(OV - \min)}{\max - \min}, \qquad (1)$$

where NV stands for a normalised value, OV represents original (nominal) value, and the difference between maximum (max) and minimum (min) is always one less than the original number of classes. The normalised values ranged from 0 to 5.

The normalised factors were used to develop the optimum landslide susceptibility model. The models were developed using the linear weighted sum [33]. The result is standardised landslide susceptibility, calculated from the eqn (2):

$$H = \sum_{j=1}^{n} w_j \times f_{ij}, \qquad (2)$$

Table 1: Weights of spatio-temporal factors of ten models used for the landslide susceptibility calculation. The "Success rate" of model is calculated from the proportion of landslides in the two lowest classes of landslide susceptibility.

MODEL	M_01	M_02	M_03	M_04	M_05	M_06	M_07	M_08	M_09	M_10	M_11_3F
Engineering-geological properties	0,3	0,3	0,25	0,4	0,1	0,166	0	0,2	0,3	0,3	0,41
Slope	0,2	0,25	0,25	0,2	0,1	0,166	0,2	0,2	0,25	0,3	0,26
Curvature	0,1	0,1	0,05	0,05	0,1	0,166	0,2	0,2	0,1	0,05	0
Aspect	0,05	0,05	0,05	0,05	0,1	0,166	0,2	0,2	0,05	0,05	0
Landcover type	0,3	0,25	0,35	0,25	0,5	0,166	0,2	0	0,1	0,25	0,33
Distance to struct. elements	0,05	0,05	0,05	0,05	0,1	0,166	0,2	0,2	0,2	0,05	0
Success rate	12,1%	10,6%	12,8%	11,3%	14%	14,3%	18,5%	19,7%	14%	13,3%	12,3%

where H represents standardised relative landslide susceptibility (0-5), w_j represents weight for a given factor and f_{ij} represents the value of continuous or discrete variable. The weight values for different factors were defined based on previous research [28] and modified or adapted to some extent by an expert decision. Altogether ten models plus one generic for the whole of Slovenia were calculated using different weight combinations (Table 1). In order to select the optimum model, a comparison of their prediction success was necessary. This comparison was based on the equal area method to avoid the differences between the model value distributions. Every model was divided into 100 classes, 1% of the area per class. The criterion for model success was the number of successive classes in which a statistically significant proportion of landslides from the test set occur. The lower the number of classes, representing the landslide susceptible area, and the higher the proportion of test landslides in the landslide susceptible area, the better the model.

Based on the analysis results a mathematical model was developed and the results represented in the form of a GIS data set and in a map. The Landslide Susceptibility Map of Slovenia at a scale of 1:250,000 is a final product of linear mathematical modelling of spatio-temporal factors that govern landslide occurrence and hence landslide susceptibility. Based on an expert decision, the areas with slopes less than 5° were excluded from the modelling and the lowest possible susceptibility was assigned to them. In areas with slopes of less than 5°, where no landslides should occur, 55 or roughly 5% of these phenomena from the testing set are present. The error is present in all of the models.

3.2 Landslide hazard assessment

The results of landslide susceptibility modelling and its spatial distribution permitted an analysis of landslide hazard distribution on a national scale. Using the landslide susceptibility model and data of spatial distribution of anthropogenic components (census, land use, infrastructure), an estimation of the hazard was performed. This cross validation enabled the assessment of component exposure to possible landslide occurrences. Although the scale is very general, the results are a very informative indicator of anthropogenic exposure to slope mass movement and an indication of whether natural processes were considered in the spatial planning process. The analyses were conducted in GIS in a 25×25 m cell. Due to the cell based analyses the results can deviate by up to 0.4% from the values shown.

4 Results and discussion

4.1 Landslide susceptibility model

The worst results in calculating the landslide susceptibility models were given by models where geological engineering properties (M_07) or landcover type were excluded (M_08). Slightly better results were achieved with the model M_06, where all of the factors were assigned equal weights. Next by performance was model M_09, where distance to structural elements was given an important role and the role of the landcover type was minimised. With model M_05 the landcover type was given a very important role and the rest was split among other factors. The success rates of the rest (models M_01, M_02, M_04, M_10) were very similar. Model M_02 was chosen for the most successful and suitable landslide susceptibility model, based on a good landslide to area ratio and the expert knowledge (and logic) of importance of spatio-temporal factors. In only 18% of area, 61.5% of landslides occur, and in less than 1/3 of area (29%), 76% of landslides occur. Split in half, to landslide susceptible and landslide "safe" areas, in landslide susceptible areas 88.2% of landslides occur. Table 2 represents the basic characteristics of model M_02. The reclassification of model M_02 values into landslide susceptibility classes, which are shown in the form of the Landslide Susceptibility Map of Slovenia at a scale of 1:250,000, are based on actual landslide occurrences compared to expected ones. In the class of highest landslide susceptibility, the areas where six times more landslides occurred than expected were classified. This class represents the top 7% of area according to landslide susceptibility, and comprises 43.3% of landslides. All areas where the proportion of landslides to that of area ratio is greater than one were joined to the class of high landslide susceptibility. 27% of landslides were located in an area of 17%. The class of medium landslide susceptibility comprises areas where the proportion of landslides to that of area ratio is near or equal to one. In this class, which spreads over 10% of the total area, 10.5% of landslides occurred. In areas with low landslide susceptibility that cover 21% of the total area, 8.5% of landslides occur, and in areas with very low, but still some landslide susceptibility, which cover 17% of the total area of Slovenia, 5.5% of landslides occur. Other areas belongs to the "landslide safe" zone where 5.1% of landslides occur. This error is most probably a result of digital elevation model generalisation (the possible generalisation of transitions between terraces) and due to the fact that analyses were conducted at a scale of 1:250,000. Cumulatively in the first class 43.3%, in the first two 70%, in the first three 80%, and in the upper four susceptibility classes 90% of landslides occur. In each of the lowest two landslide susceptible classes 5% of landslides occur.

Table 2: Distribution of landslide susceptibility class areas for model M_02. "A" represents the proportion of the area covered by a given class ("Class"). "Reclassified classes by area proportion" represents the area proportion of landslide susceptibility classes, "Model values" represents the range of model values for a given class in model M_02, "Landslide susceptibility" defines the description of susceptibility, and "Landslide proportion" states the proportion of landslides in a given class.

Class	A (%)	Reclassified classes by area proportion	Model values	Landslide susceptibility	Landslide proportion
1	28.00%	0 - 28	0 – 0.57	None	5.1%
2	17.03%	28 - 45	0.57 – 3.19	Very low	5.5%
3	20.99%	45 - 66	3.19 – 4.59	Low	8.5%
4	10.00%	66 - 76	4.59 – 5.42	Medium	11.4%
5	17.00%	76 - 93	5.42 – 7.16	High	26.2%
6	6.97%	93 - 100	7.17 – 9.88	Very high	43.3%

The analysis results indicated one particular characteristic, the importance of three spatio-temporal factors, lithological or geological engineering characteristics of rocks and soils, slope inclination, and landcover type. Using only these three factors, models would not achieve such prediction performances as in the cases presented above since the prediction would be of lower detail, but

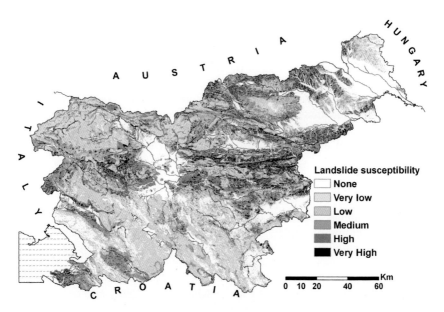

Figure 3: Landslide susceptibility map of Slovenia at a scale of 1:250,000.

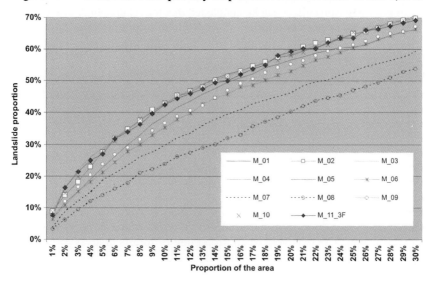

Figure 4: Cumulative distribution of landslides according to area proportion for each of eleven models. Legend represents labels for each of the linear models.

the results would still be satisfactory. The success rate analyses for these three factors revealed that the error for the ideal combination of factors in model M_11_3F (lithology, 0.41; slope inclination, 0.26; landcover, 0.33) is 12.3%, while in the chosen model M_02 the prediction error is 10.6%. The "ideal" weight values of the three factors were derived from the average of the best ten weight values for each of the factors when factor influences were analysed individually for each factor. The averaging approach was selected to reduce potential extreme variations caused by different factor weights. The comparison between model M_02 and model M_11_3F poses a reasonable question of developing complex landslide susceptibility models. This question, of course, has no solid foundations, since the quality of models is augmented with the inclusion of a reasonably higher number of spatio-temporal factors. Moreover, when the safety of inhabitants or property is concerned, every percentage point counts.

4.2 Exposure to landslides and hazard assessment

The majority, almost two thirds, of the population [34] lives in areas of neglected landslide susceptibility, while 7.8% of the population inhabits areas of very high landslide susceptibility. The difference between the proportion of inhabitants and the proportion of area is statistically significant. Almost 19% of the Slovenian population lives in areas of high or very high landslide susceptibility. It can be concluded that inhabitants do not take landslide susceptibility into consideration when planning locations for their homes.

When analysing the distribution of buildings [34] in relation to landslide susceptibility in Slovenia, it can be concluded that the relative high proportion of buildings in the upper two classes of landslide susceptibility are the result of numerous leisure homes in hilly areas (buildings with no permanent inhabitants). The total proportion of buildings in the area with high susceptibility is twice as high as the area proportion of this class. This again shows the great negligence of past processes concerning slope mass movements when choosing building locations.

The analysis of landcover type exposure to landslides was performed on the data obtained from the Ministry of Agriculture, Forestry and Food [35]. With regard to farmland, almost twice as much area is located in the most exposed class to landslides than expected. A more detailed analysis reveals that the major contributors to landslide occurrence on farmland are meadows where grass with its shallow root system acts as a poor preventive action against landslides. The proportion of forests is, as expected, lower in areas where landslide susceptibility is negligible since these areas are occupied by other landcover types (farmland, built areas, etc). The relatively small proportion of forests in areas of very high landslide susceptibility is most probably related to the protection function of the tree root system against landsliding. As expected, built-up areas have a similar distribution to building distribution. The distributions of water and open marshland are also as expected since the majority is located in areas of negligible landslide susceptibility. The small proportion of water-related landcover types in areas of high landslide susceptibility is most probably a consequence of the generalised digital elevation model. Open land with no or insignificant vegetation is mainly situated in higher or even mountainous areas where the probability of landsliding is higher, but at the same time not extremely high since these areas most probably consist of harder rocks.

Table 3: Distribution of anthropogenic components according to landslide susceptibility classes in Slovenia.

Landslide susceptibility	None	Very low	Low	Medium	High	Very high
Population proportion	65.9%	3.3%	6.4%	5.5%	11.0%	7.8%
Building proportion	53.1%	3.9%	8.4%	7.2%	15.5%	11.9%
Farmland	48.4%	5.6%	9.7%	6.8%	15.8%	13.7%
Forest and similar landcover	14%	23.8%	28.4%	12.2%	18.4%	3.2%
Built and similar sites	55.2%	4.7%	9.1%	6.8%	14.3%	10%
Open marsh land	94.5%	1.9%	1.6%	0.6%	1.1%	0.2%
Open land with no or insignif. veget.	2.31%	53.7%	30.2%	7.5%	5.7%	0.5%
Water	80.7%	5.6%	6.5%	2.3%	3.9%	1.1%
Highway	68.3%	7.2%	8.2%	4.8%	8.5%	2.9%
Fast road	73.1%	5.4%	7.2%	5.5%	7.3%	1.6%
Regional road	25.8%	16.3%	25%	10.4%	16.7%	5.9%
Local road	36.2%	7.2%	14.2%	9.6%	20.2%	12.7%
High street	89.7%	3.0%	3.2%	1.8%	1.8%	0.4%
Main city street	79.2%	3.8%	5.2%	3.7%	5.5%	2.6%
City street	78.9%	3.3%	4.7%	3.8%	6.3%	3.0%
Public path	41.74%	5.8%	11.3%	8.6%	19.3%	13.3%
Cycling path	65.3%	3.9%	6.0%	6.3%	12.1%	6.3%
Railway	60.6%	9.7%	12.6%	6.0%	8.36%	2.7%
Class area (km^2)	5675	3451.61	4254.2	2026.8	3445.5	1412.67
Class area proportion	28.0%	17.0%	21%	10.0%	17.0%	6.97%

Regarding the exposure analyses of road infrastructure [36] in Slovenia, the distributions of highways, fast roads, regional roads, high streets, main city streets and cycling paths indicate very good infrastructure planning in relation to landslide susceptibility. The distribution of regional roads is very similar to the distribution of landslide susceptibility classes, which indicates that this type of road is built regardless of landslide susceptibility. The main reason for this phenomenon is most probably the importance of these roads which connect the most distant settlements regardless of unfavourable natural conditions. The relatively high proportions of local roads and public paths in the two most exposed classes to landsliding are probably the result of their specific use (logging and transportation of wood, footpaths, etc).

The railway network is mainly situated on flat or gently sloping terrain. When crossing hilly or mountainous areas, the railway is cut into hard rock, and when crossing soft rock, they are usually protected by tunnels, entrenchments and similar constructions. The distribution of the railway network indicates that its positioning was very prudent when taking landsliding into consideration.

5 Conclusions

Landslides are the most common local geohazard problem in Slovenia. A holistic national landslide protection approach consists of several stages. In the first stage the collection of data is necessary, followed by analyses of the available data. The existence and use of the Slovenian National Landslide DataBase bring great progress to the quick response to slope mass movement threats and the field of slope mass movement prevention. Furthermore, various useful scientific results can be achieved on the basis of an analysis of slope mass movement data. The data, stored in the Landslide DataBase, serve as a basis for a better understanding of slope mass movements and assist experts in building improved prediction models of these phenomena. The data and results based on these data further serve for the production of geohazard and georisk maps, which may, with regular updating of the database, gradually improve by providing better prediction levels. Spatial mathematical modelling of natural phenomena enables the assessment of their occurrence and hence the exposure of the environment to their impact. Based on the analytical results, the legislative stage has a responsibility to conclude the circle of protection approach, but this field is following very slowly. The holistic approach does not stop at the legislative level, but involves a live, continuous process that improves with every repeated circle.

References

[1] Ribičič, M., Buser, I. & Hoblaj, R., Digital attribute / tabular landslide database of Slovenia for filed data acquisition. *Proc. Prvo slovensko posvetovanje o zemeljskih plazovih*, Rudnik živega srebra: Idrija, pp.139-153, 1994.

[2] Statistical Office of the Republic of Slovenia, http://www.stat.si/ (10/03/2008).
[3] APAT, Italian Landslides Inventory. APAT – Italian Agency for Environmental Protection and Technical Services, Department of Soil Protection and Land Resources – Italian Geological Survey, www.sinanet.apat.it/progettoiffi (4/11/2007)
[4] BRGM, Base de Données Nationale sur les Mouvements de Terrain (BDMvt), BRGM, http://www.bdmvt.net/presentation.asp (4/11/2007)
[5] Hashimoto, F., Japan's Landslide GIS in a Relational Database. *Geospatial Solutions*, **15(8)**, pp. 20-20, 2005.
[6] Liu, J. K. & Woing, T.C., A practical approach to creating a landslide database using Taiwan SPOT mosaic. *Proc. of the 20th Asian Conference on Remote Sensing (ACRS1999)*, Hong Kong, China, Asian Association of Remote Sensing, Hong Kong, pp.561-570, 1999.
[7] Evans, N. C., Huang, S. W., & King, J. P., The natural terrain landslide study, phase 1 and phase 2. Geotechnical Engineering Office, Civil Engineering Department, Hong Kong, 1997.
[8] http://www.ga.gov.au/meta/ANZCW0703003536.html#citeinfo, (02/04/2008)
[9] National Research Council, Partnerships for Reducing Landslide Risk: Assessment of the National Landslide Hazards Mitigation Strategy, The National Academies Press, Washington, D.C., p.131, 2004.
[10] http://gsc.nrcan.gc.ca/landslides/index_e.php, (02/04/2008)
[11] WP/WLI (International Geotechnical Societies=UNESCO Working Party on World Landslide Inventory), A suggested method for reporting a landslide. *Bulletin International Association for Engineering Geology*, **41**, pp. 5-12, 1990.
[12] WP/WLI (International Geotechnical Societies=UNESCO Working Party on World Landslide Inventory), A suggested method for a landslide summary. *Bulletin International Association for Engineering Geology*, **43**, pp. 101-110, 1991.
[13] Brown, W.M., Cruden, D.M. & Dennison, J.S., The Directory of the World Landslide Inventory. United States Geological Survey, Open-File Report 92-427, p.216, 1992.
[14] International Consortium on Landslides, http://www.iclhq.org/, (4/11/2007)
[15] International Consortium on Landslides, Database of Landslides of the World, http://www2.co-conv.jp/~landslide/v2/landslide/simple.php, (4/11/2007)
[16] Fajfar, D., Ravnik, D., Ribičič, M. & Komac, M., Slovenian National Landslide DataBase as a solid foundation for the landslide hazard analysis. EGU, General Assembly, Vienna, Austria, 24-29 April 2005, *Geophysical Research Abstracts*, **7**, 4 p., 2005.
[17] Ribičič, M., Komac, M., Mikoš, M., Fajfar, D., Ravnik, D., Gvozdanovič, T., Komel, P., Miklavčič, L. & Kosmatin Fras, M., Novelation and an upgade of the landslide information system and its incorporation in

GIS_UJME database: Final report. Faculty of Civil Engineering, Ljubljana, 2006, http://www.sos112.si/slo/tdocs/zem_plaz_gis_ujme.pdf
[18] Komac, M., Ribičič, M., Šinigoj, J., Krivic, M. & Kumelj, Š., Landslide occurrence analyses and landslide susceptibility map production – report, Geological Survey of Slovenia, Ljubljana, p.138, 2005.
[19] Carrara, A., Multivariate models for landslide hazard evaluation. *Mathematical Geology*, **15**, pp. 403-426, 1983.
[20] Carrara, A., Cardinali, M., Detti, R., Guzzetti, F., Pasqui, V. & Reichenbach, P., GIS techniques and statistical models in evaluating landslide hazard. *Earth Surface Processes and Landforms*, **16**, pp. 427-445, 1991.
[21] Fabbri, A.G., Chung, C.F., Cendreo, A. & Remondo, J., Is Prediction of Future Landslides Possible with a GIS? *Natural Hazards*, **30**, pp. 287-499, 2003.
[22] Crozier, M.J. & Glade, T., Landslide hazard and risk: Issues, concepts and approach. *Landslide Hazard and Risk*, ed. T. Glade, M.G. Anderson & M.J. Crozier, John Wiley & Sons: New York, pp. 1-40, 2005.
[23] Survey and Mapping Administration, InSAR DEM 25 (Digital Elevation Model). Survey and Mapping Administration, Ministry of the Environment and Spatial Planning, Ljubljana, 2000.
[24] Buser, S. (in print) Geological Map of Slovenia at scale 1 : 250.000.
[25] Komac, M., Rainstorms as a landslide-triggering factor in Slovenia. *Geologija*, **48(2)**, pp. 263-279, 2005a.
[26] ARSO, CORINE Land Cover for Slovenia 2000. Environmental Agency of the Republic of Slovenia, Ministry of the environment and spatial planning, Ljubljana, 2004.
[27] ARSO, European environmental information and monitoring network. Environmental Agency of the Republic of Slovenia, Ministry of the Environment and Spatial Planning, Ljubljana, 2005. http://eionet.elsis.si/Dokumenti/GIS/splosno
[28] Komac, M., *Landslide occurrence probability prediction with analysis of satellite images and other spatial data.* Geological Survey of Slovenia: Ljubljana, pp. 175-196, 2005b.
[29] Komac, M., A landslide susceptibility model using the analytical hierarchy process method and multivariate statistics in perialpine Slovenia. *Geomorphology*, **74(1/4)**, pp.17-28, 2006.
[30] Stančič, Z. & Veljanovski, T., Understanding Roman settlement patterns through multivariate statistics and predictive modelling. *Beyond the map*, ed. G. Lock, IOS Press: Amsterdam, pp.147-156, 2000a.
[31] Stančič, Z. & Veljanovski, T., Understanding Roman settlement patterns through multivariate statistics and predictive modelling. *Geoarchaeology of the landscapes of classical antiquity*, Stichting Babesch: Leiden, pp.179-187, 2000b.
[32] Davis, J. C., *Statistics and data analysis in geology*, John Wiley & Sons: New York, pp.80-86, 1986.

[33] Voogd, H., *Multicriteria evaluation for urban and regional planning*, Pion Ltd.: London, pp.119-121, 1983.
[34] Survey and Mapping Administration, Building cadastre. Survey and Mapping Administration, Ministry of the Environment and Spatial Planning, Ljubljana, 2005.
[35] Ministry for Agriculture, Forests and Food, Capture and farmland change monitoring 2002. Ministry for agriculture, forests and food, Ljubljana, 2002.
[36] DRSC, State road digital net. Directorate for roads, Ministry of Transport, Ljubljana, 2000.

On the definition of rainfall thresholds for diffuse landslides

L. Longoni[1], M. Papini[1], D. Arosio[2] & L. Zanzi[2]
[1]*Dipartimento di Ingegneria Idraulica, Ambientale, Infrastrutture Viarie, Rilevamento, Politecnico di Milano, Italy*
[2]*Dipartimento di Ingegneria Strutturale, Politecnico di Milano, Italy*

Abstract

Diffuse landslides may generate a lot of damage to human settlements and risks for the population because of their frequency and their diffusion on the territory. In the past, several efforts have been oriented to the definition of appropriate empirical or physical relationships to determine rainfall thresholds that can induce landslides. Usually the term "diffuse landslides" is used to indicate shallow movements such as soil slips, debris flows or landslides in porous media with a depth less than 5 meters. The objective of this work is to discuss the common empirical methods defined in the past for shallow mass movements in soils as well as to make some comments about rockfalls. In fact, the study of rainfall thresholds for rockfalls is still a challenge. To investigate the capability of rainfall thresholds to forecast landslides, three case studies are presented: Sirta and Ardenno as landslides in porous media, and Mount San Martino as a rockfall. The first application regards two towns in Valtellina, North of Italy. The available information allows us to study these areas considering also the geological peculiarity of the sites. As far as rockfall is concerned, our analysis focuses on acoustic emissions that may be related to the rupture of the rock bridges within the fractures, i.e. an initial stage of the crack propagation that may cause a future failure. The Mount San Martino case study regards a preliminary assessment considering the correlation between acoustic emissions and rainfall. Further investigations in this field are underway.

Keywords: rainfall threshold, diffuse landslides, rockfall.

1 Introduction

In the last decades, the increase of victims due to landslides over the world has led to renewing the interest in landslide thresholds as a way to reduce fatalities. Landslides can be induced by one or more factors; anyway weather condition is believed to be the factor that may cause all typologies of landslide. Several cases presented in the scientific literature show the influence of this triggering factor on mass movements. Among all weather conditions, the presence of water due to rainfall, run off and snowmelt remains one of the major causes. It is therefore of paramount importance to correlate the vulnerability of the territory with the presence of water. The importance of meteorological events will however vary with the type of mass movement. In this paper we deal with diffuse landslides. These landslides, widespread on the territory, need particular methodologies, based on simple input data, in order to be applied to extended areas. The widespread landslides (soil erosion, shallow debris flow, soil slip and some cases of rockfall), even if with lower magnitude compared to great landslides, may generate a lot of damage to human settlements. Unfortunately no defense structures or monitoring systems allow the territory to be protected from all these mass movements. Their diffusion on the territory suggests therefore a regional approach able to define alerts based on rainfall thresholds. The aim of regional early warning systems is to identify and map the risk of the territory in order to better manage the emergency of the whole territory affected by critical meteorological events. On the contrary, the study of a great landslide, defined as a local problem, can be performed with a detailed approach. Usually these hydrogeological risks are studied through an ad-hoc methodology, based on the geological peculiarity of the local problem.

The vulnerability of populations living in hazardous areas coupled to the claim for more effective emergency preparedness have led to develop precipitation thresholds. The most used alerts are based on rainfall for two reasons: first of all, as stated above, the presence of water is one of the most important triggering factors and, secondly, rainfall can be easily measured.

The concept of precipitation thresholds was introduced for the first time in 1975 by Campbell [1] and implies a correlation between intensity and duration of rainfall necessary to trigger landslide movement. Numerous authors [2–5] have developed methods (empirically and/or physically-based) to estimate the values of triggering thresholds.

The aim of our work is to apply and discuss the main rainfall threshold relationships as real time forecasting tools. For this purpose a back analysis on a particular meteorological event was investigated. Some shallow landslides were induced to move by intense precipitation occurred in Valtellina, during the autumn of 2002 (November 17–27). In Sirta and Ardenno areas, in the south of Valtellina, the rainfall caused numerous debris flows. On the basis of the available datasets, a detailed description of the applicability of the main rainfall thresholds to geological features of Ardenno and Sirta areas is presented. The comparison of real events using different relationships is aimed to understand and discuss the peculiarity of each threshold, establishing the most suitable one to forecast the hazard in the area under examination and to identify how some rainfall thresholds act according to the different geomorphological conditions. The second section of the paper regards rockfalls, that are often diffused on the rocky mountains. In almost all cases in the literature, the empirical thresholds are related to critical rainfall that induce shallow landslides, while the relationship between rainfall and rockfalls is much less investigated, despite rain is one of the main causes of rockfalls. At present, the relationship between the initiation of rockfall and rainfall intensity-duration is yet a challenge.

A comparison between precipitation measurements and evidence of microseismic activity collected by a deployed monitoring network is performed, in order to better understand the rockfall–rainfall correlation.

2 Methodological approach

The approaches used to predict landslide susceptibility range from qualitative to complex quantitative models. Hazard maps can be achieved with the application of white, black or grey box models. White box models are based on the physical relationships that lead to mass movements. The role of water is investigated through the use of hydraulic or hydrological models. Each component of the adopted model should reflect the corresponding physical laws that regulate the water action in soil and rock mass. Unfortunately, considering the complexity of the real geological assessment, each model contains some simplifications in order to solve the problem. Black box models refer to statistical or empirical analysis, where a direct correlation is required between rainfall features and mass movement, without considering the physical laws which regulate the role of water into the slope. In grey box models both statistical and physical investigations are taken into account. This paper describes rainfall thresholds that can be defined through physical (process-based) or empirical (historical, statistical threshold; [6, 7]) approaches.

Regional methods usually belong to the empirical approaches (black box models). As stated by Guzzetti *et al.* [7], the different types of empirical rainfall thresholds proposed in the last years may be classified according to: i) the extension of the investigated area (global, regional and local thresholds); ii) the type of rainfall measurement (specific for one critical event, considering also antecedent rainfalls or including hydrological conditions).

These considerations are in general suitable for the different types of movements, even if rainfall has a different role in shallow landslides or in rockfalls. In the next paragraphs a brief discussion on the relationships recommended by other authors is reported, both for shallow landslides and for diffuse rockfalls.

2.1 Rainfall thresholds for shallow landslides

Shallow diffuse landslides, despite their low magnitude in terms of velocity and involved volume, can affect urban areas causing high economic losses. Several rainfall thresholds can be used to examine a hazardous scenario involving shallow landslides. Recently some authors compared different approaches (statistic and deterministic methods) for landslide susceptibility, in order to test the most suitable relationship (e.g. [8]). The application of deterministic and statistical methods on a real case showed that the second one can provide satisfactory risk maps. The problem associated to the determinist approach concerns the simplification assumptions; moreover its poor performance depends also on the quality of the input data. This is a key issue particularly for widely spread landslides. For this grounds, in this paper we prefer to consider only a statistical approach based on rainfall as the factor causing the initiation of mass movements on a regional scale.

Shallow landslides, including soil slips, debris flows and earth movements, are strongly influenced by water circulation. Debris flows usually occur after heavy rainfall and moisture is of primary importance for soil slips. Groundwater conditions are related to rainfall through infiltration, soil features, antecedent degree of saturation and rainfall history [9]. In addition, snowmelt can increase groundwater circulation and the run off on the slope. These phenomena are very complex and still poorly understood; therefore it is common practice to simplify the problem and to try to forecast the slope behavior only by means of available and reliable data. This is the aim of rainfall thresholds. Different types of empirical rainfall thresholds for the possible initiation of shallow landslides have been proposed in the past [10].

A threshold is the minimum or maximum level of some quantity needed for a process to take place [11]. As stated previously, these methods are focused on a statistical analysis of historical data, therefore rainfall data are fundamental for this approach and only significant weather statistics allow a satisfactory rainfall threshold to be generated. The concept of rainfall threshold was introduced by Campbell [1] just for debris flows and soil slips. In keeping with this approach, many attempts have been proposed to define the minimum height of water and/or the intensity necessary to trigger landslide movements.

A review of the literature reveals that no unique set of measurements exists to characterize the rainfall parameters able to trigger slope failures. Rainfall intensity (I) and duration (D) thresholds are the most common proposed approach. Besides I-D approach, thresholds can be also based on the total event rainfall, on rainfall event (E) – duration [12] and rainfall event – intensity [2, 13]. Additionally, antecedent rainfall can as well influences groundwater level and soil moisture, leading to possible landslide initiation [14]. A simple way to consider the role of antecedent rainfall consists in establishing thresholds based on the cumulative rainfall. [4, 15–19] have correlated antecedent rainfall with the triggering of landslides. A crucial issue of this approach regards the period over which cumulative rainfall is to be considered. The choice of the time interval is strongly influenced by the geological features, the climate conditions of the investigated area along with the heterogeneity and reliability of rainfall data [20].

As well as for the antecedent rainfalls, also the choice of rainfall thresholds is related to the climatic conditions. In the following, some rainfall thresholds will be discussed in order to apply to real cases the most suitable one in terms of geological and meteorological factors.

2.2 Rainfall thresholds for rockfalls

Small-scale rockfalls are one of the major issues in mountain areas. Generally triggered by rainstorms, freeze-thaw cycles or earthquakes, rockfalls commonly occur on vertical rock slopes and cause a lot of damage in the surrounding areas. A detailed characterization of this kind of failures is yet a challenge, since it is difficult to define the fracture network that controls the stability of a rock mass. In particular, both classical and up-to-date investigation methodologies have shown to bring little knowledge about the persistence of the discontinuities [21], in terms of rock bridges connecting the opposite sides of the fractures. Therefore, it is also difficult to understand the effect of water circulation within the fracture network and, at present, no significant relationships have been proposed to relate rainfall to rockfall occurrence. In the past, direct rockfall measurements have provided qualitative evidence for a general rainfall–rockfall relation [22, 23], but no quantitative models have been developed. With respect to shallow mass movements, rockfalls seem to be induced by more moderate precipitations and the relationships defined for shallow landslides are no more appropriate. Recently some authors [24, 25] have tried with little success to explore the possibility to define quantitative relationships between rainfall and rockfall. They claim that it is fundamental to study deeply the hydro-geological features of the unstable rock mass as well as its kinematic and dynamic conditions. They also found out that information on antecedent rainfalls must be coupled to recent meteorological events in order to better understand the hydrological regime within the rock mass. The time lag between precipitation and fracture propagation (i.e., between rainfall and possible failure) is a result of the complex interaction between water circulation and discontinuity features.

In the last years, several authors have addressed rockfall forecasting by means of the acoustic emission/microseismic technique [21, 26]. This methodology seems very promising because it allows us to gain additional information on the dynamic and kinematic processes ongoing within the rock mass and related to the propagation of discontinuities. Differently from standard displacement measurements, which are usually carried out on the surface and in few locations, microseismic sensor networks are able to monitor wide areas and to record emissions generated inside the rock mass.

In Section 4, some datasets collected by a microseismic network deployed on the S. Martino Mount (Lecco, Northern Italy) will be presented. The seismic transducers, MEMS accelerometers and geophones, have been placed close to the crown of a previous rockfall, where unstable rock blocks are still in place. The analysis of seismic signal collected by the sensors is currently ongoing in order to learn to discriminate between events related to the evolution of the instability and events that can be classified as noise. In this paper we present some preliminary considerations concerning the relationship between recorded acoustic emissions and rainfall.

3 Application of rainfall thresholds to the Sirta and Ardenno case studies

Sirta and Ardenno are two small towns, located in Valtellina, in the north of Italy. A critical meteorological event occurred in 2002 (November 17–27) induced numerous shallow movements (debris flows and soil slips). We chose these two very close and comparable areas that underwent the same pluviometric event to test some rainfall thresholds. These areas have undergone different "geological history" therefore their behavior may differ. After having examined the main rainfall empirical methods generally used to define a rainfall threshold, some of them were chosen and applied to Sirta and Ardenno areas. The most suitable methodologies to the stated aim are deemed to be those defined by Cancelli and Nova [3], Caine [2] and Giannecchini [27]. The first one was chosen because its original field of application is similar to the investigated area; the second is interesting because it is considered as the universal threshold, and Giannecchini relationship is taken into account because it does not define a single minimum or maximum curve but two curves delimiting regions identifying different levels of slope stability.

3.1 Cancelli and Nova threshold

Cancelli and Nova [3] updated the work of Moser and Hohensinn [28] to adjust their relationship to the geological and climatic conditions of Valtellina. Cancelli and Nova threshold takes the form

$$Log(I) = 1.65 - 0.78 Log(D) \qquad (1)$$

The application of this relationship to the Sirta and Ardenno areas is depicted in Fig. 1. This approach provides good results for Ardenno while the analysis of Sirta events yields less reliable results.

The poor performance is evident for a longer duration. The original relationship, proposed by Moser and Hohensinn [28] for the Austrian Alps, displays similar problems. According to Moser and Hohensinn [28] these events, depicted closer to the threshold line, are those which occurred with less magnitude. Moreover, Sirta and Ardenno differ from a geological point of view. First, the value of soil saturation was higher in Sirta if compared to Ardenno because the antecedent rainfall was noteworthy in the Sirta area. Second, Sirta events depicted close to the threshold are reactivations of previous movements. Therefore, the Cancelli and Nova relationship generally has a good performance for the first activation of slope movements, but it is essential to consider the major susceptibility of the territory to reactivations. The results of the analysis performed on the Sirta dataset show that it is necessary to modify the threshold in order to consider also the state of activity of the slope.

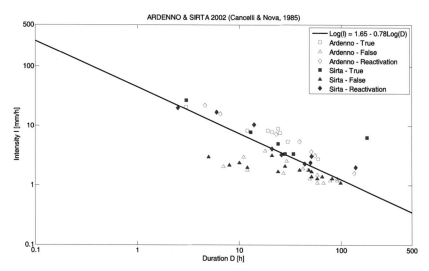

Figure 1: Application of the Cancelli and Nova threshold to the Sirta and Ardenno case studies.

3.2 Caine threshold

Caine [2] defined a worldwide rainfall threshold based on more than 73 shallow landslides where rainfall intensity and duration had been measured. Caine found that the lower bound of all plots could be expressed as

$$I = 14.82 D^{-0.39} \qquad (2)$$

The idea of Caine was to find a universal equation that could be applied to different geological, hydrogeological and topographic conditions. The relationship was intended for a general porous media, in soil debris cover. Fig. 2 shows the I-D threshold applied to Sirta and Ardenno. As in the case of the Cancelli and Nova threshold, the Caine model shows a good performance for Ardenno but not for Sirta. As stated by Caine himself, the relationship becomes less reliable for a longer duration (in our case, more than 10 hours). As a matter of fact, by comparing the Sirta and Ardenno datasets, it is clear that the Caine relationship must be updated in order to simulate appropriately the slope behavior for a longer rainfall duration. According to this, antecedent rainfalls merged with geological history (i.e., slide reactivations) of the slope must also be considered.

3.3 Giannecchini threshold

The Giannecchini threshold [27] was proposed to forecast debris flows and soil slips in the Apuan Alps (Italy). The datasets recorded across 25 years were analyzed to examine the relationships among soil movement initiation and rainfall. A threshold for shallow landslides in terms of mean intensity, duration

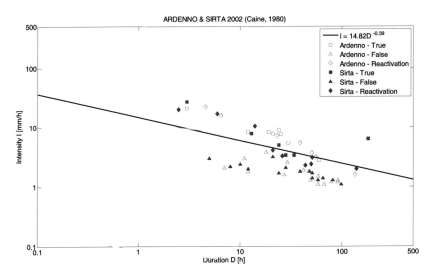

Figure 2: Application of Caine threshold to Sirta and Ardenno case studies.

and mean annual precipitation (MAP) was defined. The reliable results suggest this method can be considered valuable also for other sites. Giannecchini proposed the following threshold curves

$$\begin{cases} \begin{bmatrix} I_L = 38.363 D^{-0.743} \\ I_U = 76.199 D^{-0.6922} \end{bmatrix} for D \leq 12h \\ \begin{bmatrix} I_L = 26.871 D^{-0.638} \\ I_U = 85.584 D^{-0.7809} \end{bmatrix} for D > 12h \end{cases} \quad (3)$$

This approach is characterized by two curves, which separate fields with different degrees of stability: stability (under the lower curve), uncertain stability (between the two curves) and instability (above the upper curve).

The most important feature of this rainfall threshold regards the field of uncertainty. With regard to Fig. 3, the curves are able to discriminate Ardenno events in a proper manner, while Sirta events are closer to the lower curve. These events are the same as those discussed in the previous paragraphs.

Unfortunately, a great amount of events fall in between the two curves, which however is an area characterized by uncertainty and therefore could always be interested by false alarms.

3.4 Comparing the predictive capability of rainfall thresholds

In the previous paragraphs three rainfall thresholds, chosen in accordance with the hydrogeological features of the Sirta and Ardenno case studies, were tested. The analysis of the results shows that Ardenno events are well predicted while some errors are present when considering Sirta events. The Cancelli and Nova

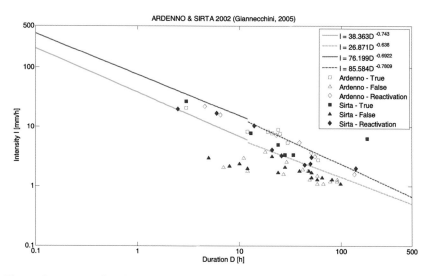

Figure 3: Application of Giannecchini threshold to Sirta and Ardenno case studies.

threshold as well as the Caine method, show inaccuracy for landslides occurring in Sirta, particularly in the case of meteorological events characterized by a long duration. As reported before, the cause may not be related just to rainfall duration, but also to the geological history of the sites. More in detail, the fact that Ardenno events characterized by a long duration are properly discriminated lead us to suppose that the key for a correct interpretation may lie in considering slope reactivation.

Different considerations must be given for the Giannecchini threshold. As reported in Fig. 3, the results of both Sirta and Ardenno true events are depicted in the region between the two curves. This region is defined as uncertain stability area. Sirta reactivations generate the same errors found using the other two thresholds. The reactivated landslides in Sirta are closer to the lower curve that represents the stability border line. Generally, for first activations or reactivations, this approach may underestimate slope stability. Thus the Giannecchini method is no more taken into account in the following.

In the literature, records regarding landslide activity show that in many cases new slides are consequent upon partial or complete reactivation of existing landslide masses. This assumption is especially true for shallow landslides. It is very common to find residual risk for this types of mass movements because of the moving back of the main scarps, the absence of vegetation that can induce a great amount of groundwater circulation, the search of a new equilibrium, the presence of the past surface of rupture that increase groundwater circulation, etc.

The site of Sirta is located in the area of a pre-existing landslide most prone to reactivation. A particular focus on reactivation must be performed in order to adapt the rainfall thresholds for these areas because the role of geomorphology on the prediction of shallow landslides is significant, but the common rainfall thresholds do not consider this factor. Generally, during rainfall events characterized by longer duration and lower intensity, the local geomorphologic features can have a significant role in influencing slope stability.

To update rainfall threshold considering also the geological history, some other cases are evaluated. Datasets regarding events in Valsassina (a valley located in the North of Italy, not far from Valtellina) induced by the 2002 meteorological event are considered. These data are particularly interesting because concerned with reactivations. Fig. 4 reports the events of Sirta, Ardenno and Valsassina along with some cases of reactivations considered by Caine [2].

With the main purpose of adapting thresholds to landslide reactivation, an updated rainfall threshold is suggested

$$I = 37D^{-0.75} \tag{4}$$

This equation is very similar to one of the relationships defined by Giannecchini. Despite that, the meaning is different: as reported in the previous paragraph, Giannecchini lower curve is the boundary condition between stability and uncertainty, while Eq. (4) represents the minimum threshold that may trigger a landslide (boundary between stability and instability).

Analyzing Fig. 4, it is possible to improve the defined threshold as a combination of two relationships: for events characterized by duration shorter than 10 hours, for safety reasons, it is possible to employ Caine threshold, while for events with longer duration it is more appropriate to consider the suggested relationship (Eq. (4)).

The use of Eq. (4) can lead to false alarms for events shorter than 10 hours, therefore it is very important to analyze the geological history of the investigated area. If the considered site was subjected to landslides in the past, it is advisable to consider the combination of Caine relationship with Eq. (4). On the contrary,

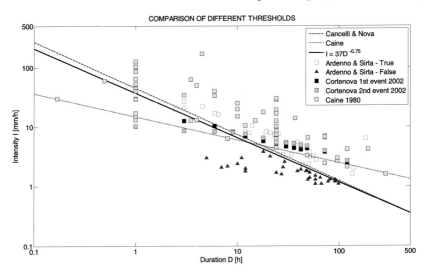

Figure 4: Comparison of rainfall threshold relationships.

if the area is subjected to first activation, it is more effective to take into account only Eq. (4) (also for duration shorter than 10 hours) in order to generate alarms correctly.

4 Mount San Martino: preliminary analysis of the role of rainfalls

A prototypal test site was chosen in the pre-alpine area to evaluate rock fall triggering mechanisms. The test site is Mount San Martino (Fig. 5), a 300-meter high rock wall threatening the city of Lecco (northern Italy). In 1969 an unexpected detachment of 15-hundred m^3 of limestone rock, destroyed several houses and caused 7 fatalities.

A microseismic network consisting of 8 Micro-Electric Mechanical System (MEMS) accelerometers, 3 geophones and 2 crackmeters was installed. Data acquisition started in May 2010 and, up to now, more than 10000 events have been recorded. The network is recording data continuously and collected events are stored in a database that can be accessed remotely at any time. To help data interpretation, which is underway, we collected meteorological parameters with a climatic station placed close to the lower part of the rock face.

Figure 5:   Mount San Martino south face (Lecco, northern Italy).

4.1 Data analysis

Preliminary interpretation of recorded datasets has been carried out cross-correlating microseismic events with precipitation. The cross-correlation is a mathematical operator that allows us to evaluate the degree of likeness between two signals. In this work, the cross-correlation has been employed to determine the time lag between acoustic emissions recorded in the time range May 2010–February 2011, and rainfall, the latter considered on an hourly and daily basis.

As in the case of shallow landslides, it is essential to take into account both recent and antecedent rainfalls. The real issue concerning cumulative rainfall regards the extent of the time period to be considered and to obtain this information we employed the cross-correlation. Helmstetter and Garambois [24] have proposed a similar approach, analyzing the correlation between rainfall and seismic signals generated by rocks falling from an unstable rock face and bouncing along a rock corridor. At the present stage, we are still not able to associate every collected acoustic emission to a distinctive physical process. According to this, we propose a preliminary analysis relating the microseismic signals to the position of the transducers, their exposure to meteorological phenomena and to the geo-morphological features of the slope. Further investigations will be carried out in order to validate these preliminary considerations. Table 1 lists the main characteristics of each sensor.

As mentioned in the previous sections, several environmental parameters can act as trigger for rockfalls. Among them, rainfall and temperature are certainly the most important and are the parameters considered here. Some periods of network down (mainly because of lack of solar energy) caused gaps in seismic monitoring and, accordingly, we were not able to perform any analysis in the months of September and December. Fig. 6 shows precipitation datasets, while Fig. 7 illustrates an example of cross-correlation between microseismic events and rainfall.

Table 1: Sensors' characteristics.

ID	Type	Position	Exposure
1	MEMS	On the unstable block	Very high
2	MEMS	On the rock face	Medium
3	Geophone	On the rock face	No
4	MEMS	On the rock face	Low
5	Geophone	On the unstable block	Very high
6	Geophone	On the rock face	No
7	MEMS	On the unstable block	High
8	MEMS	On the unstable block	No

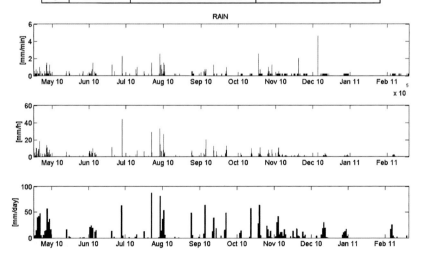

Figure 6: Rainfall from May 2010 to February 2011 measured in (top) mm/min, (middle) mm/h and (bottom) mm/day (Lecco rain gauge).

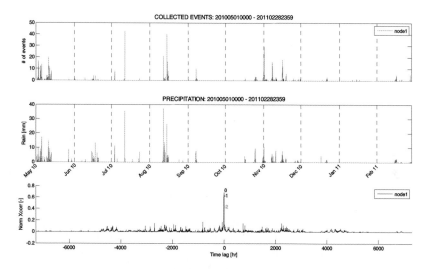

Figure 7: Example of cross-correlation between (top) hourly acoustic emission recorded by sensor 1 and (middle) hourly rainfall. The higher values of cross-correlation (bottom) occur for time lags between 0 and 2 hours.

Data reported in Table 2 express for each sensor the time lag between hourly rainfall and acoustic emissions. By comparing Tables 1 and 2, the following remarks can be pointed out:

– Sensor 1 is particularly sensitive to meteorological phenomena and we believe that, considering time lags approximately of 1 hour, acoustic emissions are mainly due to rainfall and to water run-off following precipitation.

– Sensors 6 and 8 cannot be analyzed to produce a significant interpretation because these transducers collect very few events. These sensors have been discarded in this preliminary analysis.

– The most promising sensors, because both their positions and the cross-correlation outputs are transducers 2, 3 and 4. Sensors 5 and 7 provided similar results. The fact that different sensors (MEMS accelerometers and geophones), installed in areas with dissimilar geo-morphological conditions and with different exposure to meteorological phenomena, provided comparable results can be extremely useful in order to understand the evolution of the instability.

– Sensor 3 is deemed to be the most reliable transducer. Since it is a geophone, it has better performance with respect to the MEMS accelerometers [26]. Moreover, nor rainfall neither water run-off can trigger microseismic events directly on the sensor because of its position underneath an overhanging rock.

According to the previous remarks, we decide to consider the time lags of cross-correlation between rainfall and acoustic emission collected by sensors 2,

3, 4, 5 and 7. The cross-correlation was computed considering only the months with the highest numbers of events:
- May: time lag of 20 days. In this period we have the highest value of cumulative rainfall, but particularly intense precipitations are missing. We deem that the found time lag is a plausible result as it is in accordance with the characteristics of the rainfall during the other months.
- July: three sensors display a time lag of 11 days. This month has the strongest hourly rainfall and the microseismic events have been recorded by several transducers. A shorter time lag with respect to May is in accordance with the fact that such powerful precipitations can saturate the fractures in a short amount of time and generate pressures causing the braking of the rock bridges still in place.
- November: three sensors display a time lag of about 10 days. Sensor 3 has a time lag of 24 days. We believe that in this month the effect of rainfall should be coupled to the effect of freeze-thaw cycles. Unfortunately we have no detailed information concerning the temperature within the rock mass in the areas close to the sensors since our climatic station is placed close to the lower section of the rock face. Nevertheless, it was observed that acoustic emissions mainly occurred during an abrupt decrease of the temperature that was probably close to 0°C within the rock mass. This fact, coupled to the presence of water, may have caused the water in the fractures to freeze.

Table 2: Time lags (h = hour; d = day) between hourly acoustic emission and hourly rainfall.

Month	Sensor 1	Sensor 2	Sensor 3	Sensor 4	Sensor 5	Sensor 6	Sensor 7	Sensor 8
May	1 h	21 d	-	20 d	24 d	-	-	4 d
June	-	10 d	-	-	1 d	-	-	-
July	1 h	11 d	6 d	1 h	11 d	10 h	11 d	1 h
August	1 h	-	17 d	14 h	20 h	-	2 h	3 d
October	2 h	-	11 h	1 d	17 d	-	-	-
November	1 h	10 d	24 d	10 d	8 d	-	2 h	-
January	7 d	-	10 d	4 d	20 d	-	-	-
February	1 h	-	10 d	6 d	11 d	-	-	-

The preliminary observation expressed above will be integrated with a rigorous analysis of the collected signals, extracting significant parameters both in the time and in the frequency domains. These investigations, associated to further tests performed in the laboratory and in the field, will allow us to obtain a thorough description of the physical processes.

5 Conclusions

Diffuse landslides are one of the major problems among hydrogeological hazards. These landslides, even if with low intensity, concern wide areas and therefore they are very difficult to be managed effectively. To solve this problem, the scientific community has defined some empirical relationships between mass movement and rainfall in order to robustly forecast such events. For soil slips, debris flows and shallow landslides some simple relationships have been defined. Despite that, a blind application without considering the geological peculiarities of the site is not recommended; on the contrary, it is essential to "readapt" the relationship according to the geo-morphological conditions of the investigated area. Sirta and Ardenno analysis showed the importance of tuning the found relationships on the geological context. When a detailed database of events occurred in the investigated area is available, the tuning process leads to an improved relationship for the case study. Moreover, the results of Sirta and Ardenno show the importance of the state of activity of the slope; as a matter of fact, the behavior of the slope is different if a first activation or a reactivation occurs.

Different considerations must be done for rockfall events. Forecasting rockfall is yet a challenge and these mass movements still need to be investigated more in details. Nowadays there are no empirical or statistical relationships able to define the behavior of an unstable rock face. This paper reports a preliminary study on microseismic monitoring system deployed on an unstable rock face in the North of Italy. The analysis of the San Martino test site underlines the importance of antecedent precipitations. A cross correlation in order to define the time lag between acoustic emissions recorded by each sensor and rainfall was performed. The results were examined in order to fully understand the role daily and hourly precipitation on the generation of acoustic signals. A time lag of about 20 days was determined for the most critical months, in terms of rainfall: May, July and November. This brief discussion allows only understanding the importance of the antecedent rainfall also for these types of mass movements. The use of acoustic emissions as a new monitoring approach for rockfall is yet an open problem and additional researches must be conducted in order to improve the knowledge in this field.

Acknowledgment

The authors are grateful to Matteo Negri from MeteoLecco that kindly supplied the meteorological datasets.

References

[1] Campbell, R.H., Soil slips, debris flows and rainstorms in the Santa Monica Mountains and vicinity, Southern California, U.S. Geological Survey Professional Paper, 851, 51, 1975.
[2] Caine, N., The rainfall intensity-duration control of shallow landslides and debris flows, *Geografiska Annaler*, **62A**, pp.23-27, 1980.
[3] Cancelli, A. & Nova, R., Landslides in soil and debris cover triggered by rainfall in Valtellina (Central Alps – Italy), *Proc. 4th International Conference & Filed Workshop on Landslides*, Tokyo, Japan, 267-272, 1994.

[4] Aleotti, P., A warning system for rainfall induced shallow failures, *Eng. Geology*, **72**, 247-265, 2004.

[5] Crosta, G., Regionalization of rainfall thresholds: an aid to landslide hazard evaluation, *Environmental Geology*, **35**, pp.131-145, 1998.

[6] Wieczorek, G.F. & Glade T., Climatic factors influencing occurrence of debris flows. *Debris flow hazards and related phenomena*, eds. M. Jakob, O. Hungr, Springer-Verlag: Berlin, pp. 325-362, 2005.

[7] Guzzetti, F., Peruccacci, S., Rossi, M. & Stark C., Rainfall thresholds for the initiation of landslides in central and southern Europe. *Meteorology and Atmospheric Physics*, **98**, pp. 239-267, 2007.

[8] Cervi, F., Berti, M., Borgatti, L., Ronchetti, F., Manenti, F. & Corsini A., Comparing predictive capability and deterministic methods for landslide susceptibility mapping: a case study in the northern Apennines (Reggio Emilia Province, Italy). *Landslides*, **7**, pp. 433,444, 2010.

[9] Wieczorek, G.F., Landslide triggering mechanisms. *Landslides: investigation and mitigation*, eds. A.K. Turner, R.L. Schuster, Transportation Research Board, National Research Council, Washington, DC, Special Report, pp. 76-90, 1996.

[10] Guzzetti, F., Peruccacci, S., Rossi, M. & Stark C., The rainfall intensity-duration control of shallow landslides and debris flows: an update. *Landslides*, **5**, pp. 3-17, 2008.

[11] White, I.D., Mottershead, D.N. & Harrison, J.J., *Environmental systems*, 2nd edn., Chapman & Hall: London, pp. 616, 1996.

[12] Cannon, S.H. & Ellen, S.D., Rainfall conditions for abundant debris avalanches in the San Francisco Bay region, California. *California Geology*, **38(12)**, pp. 267-272, 1985.

[13] Jibson, R.W., Debris flows in southern Puerto Rico. *Landslide processes of the eastern United States and Puerto Rico*, eds. A.P. Schultz, R.W. Jibson, Geological Society of America Special Paper 236, pp. 29-55, 1989.

[14] Glade, T., Crozier, M.J. & Smith, P., Applying probability determination to refine landslide-triggering rainfall thresholds using an empirical "Antecedent Daily Rainfall Model". *Pure Appl. Geophys.*, **157**, pp. 1059-1079, 2000.

[15] Govi, M., Mortara, G. & Sorzana, P., Eventi idrologici e frane. *Geologia Applicata*, **20(2)**, pp. 359-375, 1985.

[16] Cardinali, M., Galli, M., Guzzetti, F., Ardizzone, F., Reichenbach, P. & Bartoccini, P., Rainfall induced landslides in December 2004 in south-western Umbria, Central Italy. *Nat. Hazards Earth Syst. Sci.*, **6**, pp. 237-260, 2005.

[17] Terlien, M.T.J., The determination of statistical and deterministic hydrological landslide-triggering thresholds. *Environmental Geology*, **35**, pp. 124-130, 1998.

[18] Kim, S.K., Hong, W.P. & Kim, Y.M., Prediction of rainfall-triggered landslides in Korea. *Landslides*, 2nd edn., ed. D.H. Bell, A.A. Balkema: Rotterdam, pp. 989-994, 1991.

[19] Gabet, E.J., Burbank, D.W., Putkonen, J.K., Pratt-Sitaula, B.A. & Oiha, T., Rainfall thresholds for landsliding in the Himalayas of Nepal. *Geomorphology*, **63**, pp. 131-143, 2004.

[20] Sengupta, A., Gupta, S. & Anbarasu, K., Rainfall thresholds for the initiation of landslides al Lanta Khola in north Sikkim, India. *Nat. Hazards*, **52**, pp.31-42, 2010.

[21] Spillmann, T., Borehole radar experiments and microseismic monitoring of the unstable Randa rockslide (Switzerland). PhD dissertation no. 16866, Swiss Federal Institute of Technology, Zurich, Switzerland, pp. 205, 2007.

[22] Matsouka, N., The rate of bedrock weathering by frost action: field measurements and a predictive model. *Earth Surf. Proc. Land.*, **15**, pp. 73-90, 1990.

[23] Andrè, M.F., Holocene rockwall retreat in Svalbard, a triple-rate evolution. *Earth Surf. Proc. Land.*, **22**, pp. 423-440, 1997

[24] Helmstetter, A. & Garambois, S., Seismic monitoring of Sèchilienne rockslide (Freach Alps): analysis of seismic signals and their correlation with rainfalls. *J. Geophys. Res.*, **115**, F03016, 2010.

[25] Paronuzzi, P. & Gnech, D., Frane di crollo indotte da piogge intense: la casistica del Friuli Venezia Giulia. *Giornale di geologica Applicata*, **6**, pp. 55-64, 2007.

[26] Arosio, D., Longoni, L., Papini, M., Scaioni, M., Zanzi, L. & Alba, M., Towards rockfall forecasting through observing deformations and listening to microseismic emissions. *Nat. Hazards Earth Syst. Sci.*, **9**, pp. 1119-1131, 2009.

[27] Giannecchini R., Relationship between rainfall and shallow landslides in southern Apuan Alps (Italy). *Nat. Hazards Earth Syst. Sci.*, **6**, pp. 357-364, 2006.

[28] Moser, M. & Hohensinn, F., Geotechnical aspects of soil slips in Alpine regions. *Engineering Geology*, **19**, pp. 185-211, 1983.

Computer analysis of slope failure and landslide processes caused by water

I. Sarafis[1] & J. Zezulak[2]
[1]*Mechanical Engineering Dept., T E I, Kavala, Greece*
[2]*Dept. of Constructions, Czech University of Agriculture, Prague, Czech Republic*

Abstract

At headwater channels in particular, the landslide of slopes with water body intersection, together with earth dam failures frequently become a main contribution to debris flow. The use of a computer model for slope stability evaluation of an arbitrary earthwork based on the Bishop's method is discussed. Similarly to that of Peterson, it adopts the assumption of a circular shape of the sliding body and evaluates the balance between the forces acting in the direction of the failure (destructing forces) and the resisting forces. In contrast to Peterson, the Bishop's method considers solely the 'mobilised' shear stress as the only part of total shear stress over the sliding circle. The program applies both for earth dams and for natural land-sliding. The AutoCAD engine provides for visual generation of an almost unlimited number of shapes of the body 'segments', being interfaced with the input/output utilities of the soil mechanics module.
Keywords: earth bodies, landslides affected by water, slip circle, CAD support.

1 Introduction

Apart from methods of finite elements and/or differences, the traditional slope stability solvers are classified into two categories, using either principles of elastic theory or conditions of boundary plastic balances. At the latter, the limiting state of sliding body is assumed and conditions of its genesis are examined. The correct presumption of the form and position of slipway surface is a crucial point of such analysis. Different authors consider the problem differently.

Possibility of an easy introduction of various types of soils into the earthwork and introducing features of non-homogeneity into its structure becomes main advantage of the slip-circle oriented methods. Most often assumed cylindrical, though composition of several planar elements, wedge-shaped or entire block soil sliding can also simulate processes in a slipping zone. One can consider the form of lemniscates, logarithmic curves, spirals but the circles still prevail. An experience has proven that, no matter how large the radius could be, the presumption of circular cylindrical surface fits well both to forming the failure and to simplicity of algorithmic structure. This is why it still prevails in many methods of slope stability determination such as the method of Peterson and, more recently that of Bishop [3]. A reader with deeper interest in theoretical basis of the algorithm is kindly referred to the program manual, Novotny [1], Zezulak and Novotny [2].

2 Mathematical formulation and algorithmic structure

2.1 Bishop's method: principles

After Bishop, solely the mobilized portion of the shear stress is introduced into inactive forces' calculation, as the only part of its full value,

$$\tau_f = \frac{1}{F}\left[c' + (\sigma - u)\,\mathrm{tg}\varphi'\right] \quad (1)$$

where τ_f shear stress
  F resulting safety factor
  c' effective soil cohesion along the slip circle
  φ effective angle of friction along the slip circle
  σ normal tension actuating upon slip surface

As the safety factor is not known beforehand its evaluation requires an iterative procedure. We write an equilibrium condition at the centre of an investigated slip circle. For the value of σ we introduce $\sigma = P/l$. For active and passive moments with respect to the centre of a circle it follows, Figure 1:

$$M_a = \sum Wx$$
$$M_b = R\sum\left(c'l + P'\mathrm{tg}\varphi'\right) \quad (2)$$

If considering the differences in forces E_i, E_{i+1} and X_i, X_{i+1} then the force P' can be evaluated from cumulative condition written in vertical direction, Figure 1:

$$P' = \frac{W + X_i - X_{i+1} - l(u\cos\alpha + \frac{c'}{F}\sin\alpha)}{\cos\alpha + \frac{\mathrm{tg}\varphi'\sin\alpha}{F}} \quad (3)$$

In view of further analysis let now introduce $X\text{-}R\sin\alpha$, hence $R=X/\sin\alpha$. The coefficient of Bishop $\overline{B}$ follows from

$$u = \overline{B}\left(\frac{W}{b}\right) \quad (4)$$

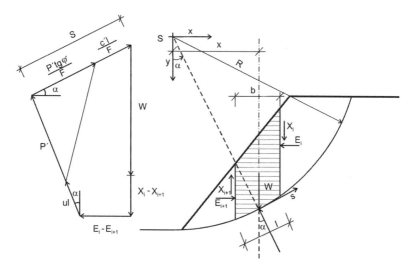

Figure 1: Equilibrium of active and passive forces.

It is called a global pore pressure coefficient, defined as

$$\bar{B} = \frac{\Delta u}{\Delta \sigma} = \frac{\text{pore pressure increment}}{\text{max tension increment}} \quad (5)$$

The $\bar{B}$ coefficient should be laboratory-determined for load reflecting the real-life conditions. We express the safety degree as a rate of passive and active moments. After substitution of the above expression (3) for P', we receive, after rearranging the original implicit expression for the safety degree F after Bishop,

$$F = \frac{M_a}{M_b} = \frac{R\sum(c'l + P'\text{tg}\varphi')}{\sum Wx} =$$
$$= \frac{1}{\sum W \sin\alpha} \sum \left[c'b + \text{tg}\varphi'\left(W(1-\bar{B}) + X_i - X_{i+1}\right)\right] \frac{1}{\cos\alpha + \frac{\text{tg}\varphi' \sin\alpha}{F}} \quad (6)$$

According to Taylor [4], and in spite the fact that forces at boundaries of elementary strips do not vanish, the overall forces' equilibrium is preserved. One can suppose $\Sigma(X_i - X_{i-1}) = 0$ with sufficient accuracy. Then the eqn. (6) can be further simplified.

2.2 Pore pressure analysis

In water saturated sector of the earth-body the pore pressure can be evaluated by introduction of the factor r_u into governing equation,

$$r_u = u/w^\times \quad (7)$$

where r_u dimensionless factor
       u neutral tension (equal to hydrostatic pressure)
       $w^\times$ tension from weight overburden considered at neutral tension pressure centroid

The distribution of the factor r_u follows from orthogonal flow net. From forces' balance the following can be worked out, in analogy to its original (6),

$$F = \frac{1}{\sum W \sin \alpha} \sum [c'b + W(1-r_u)\operatorname{tg}\varphi'] \frac{1}{\cos\alpha + \dfrac{\operatorname{tg}\varphi' \sin\alpha}{F}} \qquad (8)$$

2.3 Governing equation algorithmic structure

Here we start again from the eqn. (6) that will be adopted into an algorithmic form necessary to accomplish: arbitrarily torn up ground, non-homogenous structure of the earth body cross section, arbitrary composition of soils of different properties, introduction of hydrodynamic pressure caused by seepage water, slope stability affected by earthquake and, external load upon the crest.

2.3.1 Broken surface and non-homogenous structure of the earth body

If the soils within various segments are homogenous, for each elementary strip dx it applies $dW = \gamma_z h(x)dx$ provided the function $h(x)$ of the strip height is smooth and continuous.

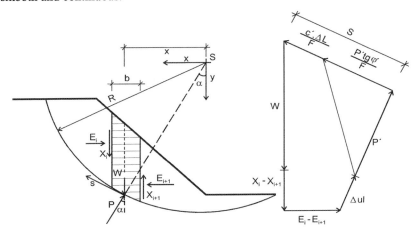

Figure 2: Equilibrium of active and passive forces, pore pressure included.

The γ_z is a specific weight of the z-indexed soil. The elementary strips dh can be integrated throughout entire region Γ of the cross section. After rearrangement, the safety factor is computed with the equation, Figure 3:

$$F \frac{\sum\limits_\Gamma \int\limits_{x_i}^{x_j} [c' + \operatorname{tg}\varphi' \gamma_z h(x)(1-\bar{B})] \dfrac{1}{\cos\alpha + \dfrac{\operatorname{tg}\varphi' \sin\alpha}{F}} dx}{\sum\limits_\Gamma \int\limits_{x_i}^{x_j} [c' + \gamma_z h(x) \sin\alpha] dx} \qquad (9)$$

Non-homogenous earthworks can develop into quite complex formation and usually require CAD support for its visual definition. Here we apply the following procedure in computerization: (1) first we specify types of soil that differ in physical-mechanical properties and introduce co-ordinate system so that its origin coincides with the centre of the slip circle, (2) we transform all the geometric and soil layer data hereinto, (3) for sake of numerical integration we subdivide the segment above considered surface into elementary pinstripes.

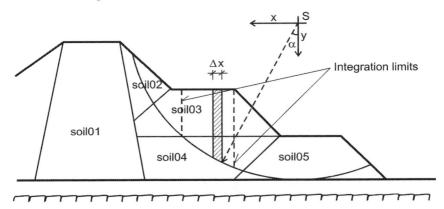

Figure 3:   Definition sketch of non-homogenous earth dam.

Any two breaking points on earthwork surface, soil transitions and/or slip circle intersect define a limit of integration, see Figure 4: for more details. We arrive to interim form of governing equation, yet missing the impact of earthquake,

$$F = \frac{\sum_\Gamma \int_{x_i}^{x_j} \left[c' + \mathrm{tg}\varphi' \gamma_z \Delta W \left(1 - \overline{B}\right)\right] \frac{FR}{F\sqrt{R^2 - x^2} + \mathrm{tg}\varphi' x} dx}{\sum_\Gamma \int_{x_i}^{x_j} \frac{x}{R}(\Delta W) dx} \quad (10)$$

$$\Delta W = \left[A_R + B_R + (D_R - C_R)(x_i - x) + \gamma_1 \sqrt{R^2 - x^2}\right]$$

where A_R, B_R, C_R and D_R are auxiliary coefficients [2].

2.3.2 Earthquake impacted slope stability

In earthquake prone regions particular attention should be given to the slope stability. Provided its impact is proportional to the soil specific weight and it acts horizontally, it can be understood as an internal body force, $Z = \xi \gamma_z$, where ξ is a coefficient proportional to the quake intensity,

$$F = \frac{\sum_\Gamma \int_{x_i}^{x_j} \left[c' + \mathrm{tg}\varphi' \gamma_z \Delta W \left(1 - \xi \frac{x}{R}\right) - u\right] \frac{FR}{F\sqrt{R^2 - x^2} + \mathrm{tg}\varphi' x} dx}{\sum_\Gamma \int_{x_i}^{x_j} (\Delta W) \left[\frac{x}{R} + \xi \frac{\sqrt{R^2 - x^2}}{R}\right] dx} \quad (11)$$

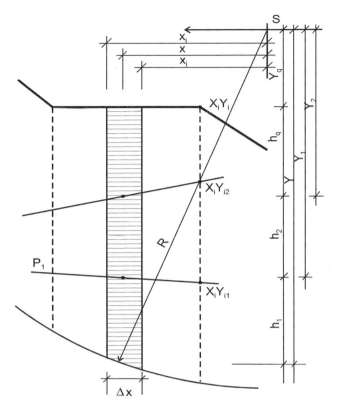

Figure 4:   Elementary strip weight definition.

2.3.3 Complete formulation of the governing equation
If compared with original eqn. (6) of Bishop, the eqn. (11) conforms with all additional impacts, such as arbitrary form of ground, non-homogenous structure of the earth body, arbitrary composition of soils of different properties, introduction of hydrodynamic pressure caused by seepage water, slope stability affected by earthquake and external load over the crest.

2.3.4 Field of application
The program solves stability of an arbitrary geometry of earth body for maximum of 14 types of soils in the construction/natural hill slope. The number of geometry-definition lines (called the 'segments') is unlimited. The program works in a man-machine conversation control and user-defined data files. The import/export of the data from/into a CAD system is supported.

The program is implemented within design offices throughout the Czech Republic and also in several UN projects (Bangladesh, Burma). All the data processing and the draftsmen work should be preferably CAD computerized. On the output the program provides for the critical slip-circle definition and corresponding minimum safety degree. Its postprocessor can also evaluate total bulk of soil entering the stream.

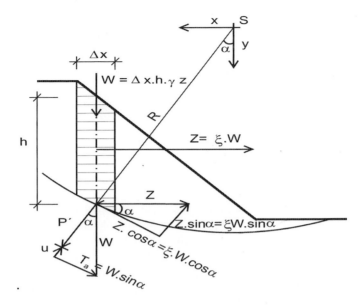

Figure 5: Resolution of forces impacted by earthquake.

2.3.5 Program requirements
(1) Given geometry of the earthwork and distribution of individual soil types within the earth body. The Cartesian coordinate system x-horizontal, y-vertical.
(2) Pre-processed (or estimated) phreatic surface of groundwater table creates boundary between the dry and the wet soil.
(3) The following soil parameters in all segments of the construction:
 -Specific weight of water (γ-water) (kN.m^{-3}).
 -Specific weight of individual soil types (γ_z-soil) (kN.m^{-3}).
 -Angles of internal friction of the soils φ'.
 -Cohesion factors of the soils c' (kN.m^{-3}).
 -Bishop's coefficients.
(4) Constraint for computation process in form of a boundary polyline, beyond which no circles can be generated.
(5) Value of external load acting upon the construction, if any.

2.3.6 Selection of computational strategy
The program Soil_SlipCircle provides for three options in minimum safety factor analysis:
(1) **Single circle:** given fixed centre and radius R.
(2) **Set of circles:** given fixed centre and increment in R.
(3) **Set of circles:** given initial centre, increments in R, and in x and y coordinates.

The strategies vary with respect to complexity of calculations. In the first case only a single slip circle is computed while given its centre and radius. The second case provides a set of circles from a given centre and increment in the radius. The third case creates all possible circles and it is a most effecting option in defining the safety factor. The program terminates the search when solution circle hits constraining boundary.

2.3.7 Input data
(1) **Earthwork geometry:** SegmNo, x_{start}, y_{start}, x_{end}, y_{end}, SoilNo, ExtLoad available for each geometry-construction line/polyline of the earthwork. The text-form data can be replaced by DXF entry using AutoCAD.
(2) **Soil parameters:** FileHeader, JobDescriptor, No of the 1st wetted soil from the top, Earthquake coefficient (RichterScale), SoilNo and related parameters (specific weight γ_z, internal friction angle φ', soil cohesion factor c', Bishop's coefficient B).
(3) **Man-machine conversation:** choice of geometry data input from the text file or from the DXF-file (e.g. from AutoCAD), selection of the computational strategy (3 options). The program controls the syntax of entered data and issues necessary warnings.

2.3.8 Output data
(1) **Text output:** program generates the detailed review of input data: soil parameters, strategies used in search for the minimum safety factor, starting conditions included and results of the analysis: list of all circles when safety factor $F < 3.0$, minimum safety factor F found, statistics of the results (number of circles computed, elapsed computer time, etc.).
(2) **Visual output:** the earthwork geometry and a critical slip circle can be viewed by any CAD system with DXF interface.

2.4 Selected examples and case studies

Two examples are provided to demonstrate the Soil_SlipCircle functioning for non-homogenous earth dams stability evaluation:
(1) **Primitive case:** soils layered horizontally.
(2) **Complex case:** soils layered with respect to dam building technology.

2.4.1 Example of the dam, soils deposited horizontally
This is a simple case to demonstrate virtue of the geometry definition using the CAD system. The earthwork consists of 5 horizontally layered soils of various physical-mechanical properties. Computational strategy is set to option 2, see 2.3.6, i.e. set of circles is given of fixed centre and increment in R, see Figure 6:

2.4.2 Soils deposited with respect to the dam building technology
The case study of the dust and soot-collecting reservoir reflects the technology of gradual filling of the pool by thermal power plant soot and dust. Although obsolete, it may serve an example for natural unstable hill slopes stability analysis. The case consists of 115 lines created in AutoCAD, and of 12 different soil categories. The program has evaluated more than 1600 events, to achieve the value of minimum safety degree $F=1.694$, Figure 7:

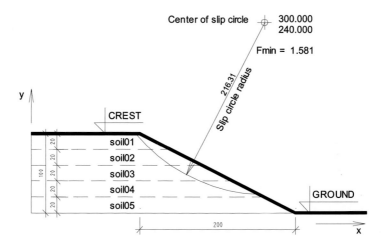

Figure 6:   Example of 5 horizontally layered soils.

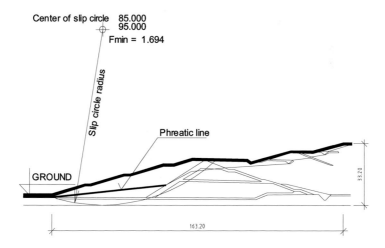

Figure 7:   Example of complex earthwork system.

3   Conclusions

The program Soil_SlipCircle evaluates the slope stability of non-homogeneous earth bodies by modified method of Bishop. Its main features are as follows:
    (1) The slip circles are generated in three program options:
        (a) Given the radius and the center.
        (b) Given the center and increment of the radius.
        (c) Given the set of centers, increments in X, Y and in the radius.

(2) Construction of the constraining polyline that prevents circle abnormal generation.
(3) Inconsistency checking of geometry data on input.
(4) Introduction of earthquake- and Bishop's coefficients and of external load upon the ground.
(5) Two options in the soil water saturation: dry soil, phreatic surface

References

[1] Novotny, R., *Program Bishop, Slope stability evaluation (in Czech)*, Energoprojekt Praha, Ltd., 1992.
[2] Zezulak, J., Novotny, R., *Program Soil_SlipCircle, Theory and Users' Guide*, Czech Univ. of Agricult. Dept. of Constructions, Prague, 2006.
[3] Bishop, A. W., Morgenstern, N., *Stability Coefficients for Earth Slopes*, Geotechnique X 4 pp.129-150.
[4] Taylor, D., W., *Fundamentals of Soil Mechanics*, New York, 1948.
[5] Mencl, V., *Soil and rock mechanics (in Czech)*, Academia Praha, 1966.
[6] Bazant, Z., *Foundation Engineering (in Czech)*, Academia Praha, 1973.

The integration between planning instruments and evaluation tools in the management of landslide risk

M. Magoni
DiAP, Politecnico di Milano, Italy

Abstract

The paper shows the main factors to integrate the planning instruments, the evaluation tools, and the certification procedures for the management of landslide risk. The integration is based on the Risk Analysis for the methodological aspects and on the Territorial Risk Certification for the procedural ones. The Risk Analysis structures the evaluations in three main factors: the quantity of goods, people and activities with different levels of sensitivity (exposition), the probability that goods, people and activities could have damage (vulnerability), and the nature, the frequency and intensity of potential catastrophic events (hazard). The Territorial Risk Certification structures the procedure in four main steps: territorial and risk analysis; the production of a structure of objectives and targets; the individuation of strategies and the planning of interventions; the monitoring of territorial risks factors and the adoption of corrective measures.

Keywords: instruments of landslide risk management, landslide risk evaluation.

1 Introduction

Many of the landslide events, like all the catastrophic events, differ from ordinary phenomena because of their elevated drama due to the speed of their arrival, the difficulty in foreseeing the time and intensity of their occurrence and the large amount of damage which is produced in a short time. For example, the effects of a landslide can ruin buildings or make them collapse, much more quickly than the effects that time has on buildings.

In order to limit damage due to landslide events, man tries to intervene whenever possible on the processes that cause a disaster as well as the related potential and real effects. Furthermore, experts elaborate, refine and utilize specific methods of analysis and evaluation designed to improve the efficacy and efficiency of actions to prevent and mitigate catastrophes. These methods, which range from the catastrophic event tree to the analysis of risk, permit evaluation of the totality of the causal effects which underlie a landslide, with varying degrees of depth and systematicness. These can then support current policies of prevention and mitigation of territorial risks, which are increasingly finalized at attaining territorial conditions which maintain an acceptable level of risk over time, whose acceptability threshold tends to decrease as the economic well-being and perception and conscience of risks increases in the population.

The intent of this article is to furnish a contribution in order to arrive at an increased integration between planning instruments, evaluation tools and certification procedures in the management of landslide risk. This integration, which should be carried out in both procedural and methodological terms, constitutes an important step in order to improve the capacity of elaboration and implementation of interventions for landslide prevention and mitigation.

2 The instruments for the reduction of landslide risk

Interventions for reducing landslide risk can be distinguished with respect to their varying types:

- form, such as the financing of the construction of regimation works or the approval of urbanistic norms;
- object of interest, which can regard factors causing a calamitous event or its targets;
- structural nature, when they act upon the cause, such as the containment of the processes of deforestation like the prohibition of building on areas of landslide;
- mitigating, when they act on the effects, for example, building the retaining and consolidation walls of landslides;
- and emergency, when they monitor the movements of a landslide rather than limit the negative effects following a catastrophe which has just occurred.

The majority of instruments of intervention in this sector fall under the direct or indirect competence of the planning instruments, evaluation tools and certification procedures.

Planning instruments have an effect on the character of various types of settlements, infrastructures and plants, on where they are located and on the methods with which they are carried out. The plans can then act on the causes, for example avoiding the placement of a road infrastructure along an area of landslide; or on the effects, for example prohibiting construction in areas subject to landslide.

The methods with which planning tools can affect landslide risk are a function of the scale of intervention, which can range from the city to the region, and their general, sectoral or actuative nature. In this way, planning tools can act on one, more, or all factors which could create damages on things or people as a result of a landslide, in order to prevent or mitigate the relative effects.

If the general plans determine and orient the major transformations of a territory, and therefore define the relationship between possible catastrophic events and human activities and settlements, sectoral plans define the strategies and program interventions designed to resolve problems of a specific plan of action. They are thus very different among themselves because they can affect industrial activity or road infrastructure, refuse or energy. Among the sectoral plans are those which specifically relate to the management of risk, such as the plans of the hydrographic basin, emergency plans, and risk prevention plans.

Actuative plans and programs, whose function is to determine the methods and timing of transformation of those portions of the territory most interested by the dynamics of development, are particularly useful tools needed in order to ensure that the activities of the above plans are correctly carried out.

Instruments of environmental evaluation are used to structure a plan's decision making process, in order to improve the quality of the solutions and increase consensus on the available choices, and to verify that the choices made correspond to the needs of the evaluation. These instruments fall in two categories: those which evaluate the project, and those which evaluate the plans and programs. The former evaluate the environmental impact, and are mature tools since they have been systematically applied for around 15 years. The latter is a strategic environmental evaluation, and is in the consolidation phase, having been introduced by a European directive in the year 2001. Both tools are divided in three phases:

- the ex-ante evaluation, which supports the preparation of the project or plan;
- the in itinere evaluation, which is the verification of the progress of the project;
- the ex-post evaluation, which verifies the intervention performed or the plan carried out.

Through the certification of territorial risks, it is possible to consider a large part of the direct and derived interventions that influence landslides, since this requires that a public administration instigate not only direct actions, meaning those for which it is directly responsible, but also indirect ones. These are the responsibility of those who are motivated, when necessary, to adopt practices for reducing landslide risks in order to maintain a relationship with the public administration. It could be possible to certify those companies which operate in hazardous areas.

Furthermore, since certification procedures verify the real achievement of objectives, their adoption will tend to favor the planning of interventions in line with available financial resources and capacity of intervention.

If the systematic verification of the work carried out by the public administration, as well as the results of the totality of the tools which this uses to reduce territorial risk, including the landslides, creates a common aspect with the evaluation instrument, that aspect specifically characterizing the certification is furnished by the search for a continual improvement of the conditions of territorial, and landslide, risk.

Since evaluation tools should be integrated with planning tools (otherwise they are useless), both tools should be regarded as the only and the widest ones to use for planning, which we can define as integrated planning tools. The integration between these two tools means that the latter uses a strategic approach, which entails the adoption of objectives which are verifiable over time and therefore measurable. The adoption of strategic planning approaches is destined to become increasingly more widespread, following the receipt of Directive 42/2004/CE about the VAS by the Member States.

Integrated planning instruments are part of the set of tools that an administration can utilize to reduce territorial risks, once it decides to certify itself, and therefore they make up fundamental reference points for certification. On the other hand, certification favors the achievement of the objectives to reduce territorial risks assumed in the plan since they try to ensure a coherent behaviour of the entire public administration regarding those objectives and verifies the correct determination and execution of the actions indicated by the general, sectoral and executional plans falling under the authority of the public administration.

The phases of certification comply with those phases that characterize the planning processes when the latter adopt strategic approaches; since certification adopts more structured procedures than do planning tools, in order to integrate the two instruments, it is preferable to maintain the certification procedures as a reference.

3 The references for the integration of planning, evaluation and certification tools

The procedure for the management and certification of territorial risks outlined in the European research project Quater (Quarter [5]) was structured in a set of passages which, in order to improve their handling and integrate them with the procedures of integrated planning, are grouped into the following phases: analysis of the territory and the relative risk factors; the determination of the strategies and the programming of the intervention actions; and monitoring the factors of territorial risk and the adoption of corrective measures. It is understood that the process of execution of the total of the interventions of territorial risk, including landslide risk, indicated by the various instruments requires a continual interaction among the instruments, and within them, among the different phases with which they are carried out.

3.1 The analysis of landslide risk factors

The analysis of landslide risk factors has the dual objective of quantifying the risk levels of a territory and determining the possibilities of intervention for their reduction. This should also consider the interaction between one or more risks, for example between those of earthquakes and landslides.

This phase is supported by methods, such as the risk analysis; the version presented in the Quater research is well suited to integrate the elaboration of the planning, evaluation and certification tools.

The risk analysis allows support not only of the analyses, the evaluations and the verifications relative to catastrophic phenomena due to landslides, but also the determination of preventive and mitigating interventions and the evaluation of their efficacy and cost-effectiveness (if economic evaluations are performed).

The risk analysis can easily be integrated with multi-dimensional integration methods, both aspatial types (multi-criteria analysis, hierarchic analysis, multi-decisional analysis and for certain aspects cost-benefit analysis, so that effects of alternative solutions may be compared) as well as territorial types (such as cluster analysis and land suitability analysis) in order to contribute to the classification of a territory with respect to levels of localization or intervention criticality.

The risk analysis structures the analyses in 3 factors which make up the references of the evaluation methods of the single risks:

- exposition (the presence and quantity of goods, the number of people and activities with different levels of sensitivity);
- vulnerability (the probability that goods, people and activities can be damaged);
- and hazard (the frequency and intensity of potential catastrophic events, including the landslides).

However, it is not applied in the same way for all types of risks, since the latter has different characteristics and requires specific approaches. For example, the hazard from landslides can be evaluated, in terms of the probability that a certain event occurs in a given location in a given time interval, similarly to the hazard of earthquakes. However, since water flows are able to be controlled, the catastrophic events deriving from floods, as opposed to those from earthquakes, can be partially controlled by diverting waters to less vulnerable areas. Furthermore, the hazard of industrial accidents, as opposed to the hazards from landslides and earthquakes, is linked to the decision of whether or not to make certain industrial activities, to their localization and to the plants and substances treated, which makes it possible to prevent these types of events.

3.2 The definition of objectives

The definition of the objectives must proceed in a structured way and must be translated in targets, which are objectives measurable over time. This passage is fundamental in order to carry out the systematic verification of the execution of the interventions.

For all types of risks, the articulation of the objectives and related targets can follow certain criteria. For example, the table 1 outlines the goals determined in the Territorial Plan of the Province of Cremona related to the objective of containment of the criticality due to territorial risks.

3.3 The determination of the strategies and the programming of interventions

The political-programming phases of certification are analogous to those of the planning processes, even though in the former case one needs to consider all instruments available to the certifying authority, among which the planning instruments.

Table 1: Indicators for the objective of the containment of territorial risks from the Territorial Plan of the Province of Cremona.

Targets:	Maintain current levels of flood risk
Indicators:	Areas of settlements per level of flood hazard; Areas of settlement for level of flood risk
Targets:	Maintain current levels of industrial risk
Indicators:	Area of settlement per level of industrial hazard; Area of industrial settlement at risk for large accidents per level of territorial vulnerability; Area of settlements per level of industrial risk
Targets:	Maintain current levels of landslide risk
Indicators:	Number of buildings subject to the landslide risk

In this phase the intervention strategies must first be determined and then determine and plan the actions needed to carry out these strategies.

The determination of the strategies must take into consideration the objectives of security of the territory assumed by the public administration, and by the need to constantly improve the risk levels, and must propose solutions which integrate the territorial, social and economic dimensions of the problems. In this phase, it is often necessary to compare different alternative strategies of risk reduction, which implicates the use of multi-dimensional analysis methods and the construction of scenarios, with a medium to long period as a reference point with respect to the desired transformations.

The objective of programming the actions of intervention is to achieve the objectives of territorial security and requires the determination of the set of human, organizational and financial resources necessary to arrive at the goals.

Reference to the programs of management of territorial risks, including landslide risk, are a valid support in general, sectoral and executional planning, while the management of the emergency phases, which differs from classical planning instruments, finds that the territorial and urbanistic plans are the tools needed to realize those conditions which permit them to operate effectively in such situations.

3.4 The monitoring of landslide risk factors and the adoption of corrective measures

Certification requires the management of the execution of the total interventions indicated for the reduction of landslide and territorial risks. To this end, an audit phase aimed at controlling the efficacy of the strategies elaborate is made. The audit evaluates the retroactions that go from the Territorial Risk Management System to the program of interventions, and determines the corrective actions necessary to make to the management systems and eventually the modifications to make to the strategies and objectives

The monitoring and the consequent adoption of corrective measures is an activity which doesn't fall under traditional planning procedures but which is more widely used with the strategic environmental evaluation, of which the in itinere evaluation is a fundamental part of its execution.

The monitoring of landslide and territorial risks assumes a specific character with respect to other types of monitoring since it is necessary to carry out the correct execution of the interventions, both in periods in which there are no calamitous events (in which case the correspondence between the prevention and execution of the interventions) as well as in the periods following catastrophic events. In this case, previsions made on the vulnerability and on the exposition and the efficacy of the interventions of prevention and mitigation can be verified.

4 Conclusions

The integration of planning, evaluation and certification tools enables improvement of the efficacy of the interventions of landslide risk containment. This integration should take into consideration the fact that the content of these tools can be characterized by their aspects of emergency, which are closely linked to the management of calamities when they arise, or structural, which serve to carry out the conditions needed to reduce the risk, in order to render the integration stronger. In particular, it is important to structure the pathway that goes from the management of the emergency, made through the Emergency Plan, to the mitigation of the impacts and the realization of structural interventions, operations that are carried out using the Risk Prevention Plan and the implementation in the general plans and the sectors of action and projects to improve safety. In this way, structural indications are connected with the strategic-structural visions of the plans, through which it is possible to outline scenarios of reference.

The graduality of the method developed under the territorial risk analysis, including landslides, which calls for three phases of successive analysis with an increasingly higher degree of investigation, is suitable for use in integrated certification and planning, since both instruments can postpone the decisional refinements to successive phases. Furthermore, the gradual increase in the analysis enables the realization of important cognitive synergies among the various instruments.

The hoped-for widespread diffusion of instruments for the management of landslide risk goes from a simplification of the related methods and procedures in order to limit the time and costs of studies resulting in a greater efficacy in the actions. To this end, the integration of these instruments provides a significant contribution following the simplification of the procedures, the standardization of the various approaches and the reduction of time dedicated to the collection and elaboration of data.

If it is easy to achieve the integration of the procedures between planning and certification, the methodologic integration is not completely resolved, since a common scale of evaluation still has not been determined on which to make the elaboration for different risks convene. This passage creates an element of important interest under the operational viewpoint since it would facilitate the attribution of the priorities for the realization of the set of the interventions of the various territorial risks enabling the achievement of aggregate or however comparable evaluations.

References

[1] Australian Geomechanics Society, 2000, *Landslide risk management concepts and guidelines*, Australian Geomechanics, March 35, 49-92.
[2] Casini L, Bonnard C H, Corominas J, Jibson R, 2005, Landslide hazard and risk zoning for urban planning and development, in: Hungr O, Fell R, Couture R, and Eberthardt (Eds), *Landslide risk management,* Taylor and Francis, London, 199-235.
[3] Crozier M J, 1993, *Management issues arising from landslides and related activity*, New Zealand Geographer 49, 35-37
[4] Glade T, Anderson M, and Crozier M J, 2005, *Landslide Hazard and Risk*, Wiley, Chichester.
[5] QUATER, 2004, Territorial Risk Management Systems of Municipality", Project INTERREG IIIB MEDOC, Research group of Politecnico di Milano coordinated by Maria Cristina Treu.

Finite element analysis of the stability of artificial slopes reinforced by roots

F. Gentile[1], G. Elia[2] & R. Elia[1]
[1]*Dept. of Agro-Environmental and Territorial Sciences (DISAAT), University of Bari, Italy*
[2]*School of Civil Engineering and Geosciences (CeG), Newcastle University, UK*

Abstract

The paper deals with the assessment of vegetation contribution to slope stability, with particular emphasis on the mechanical effects provided by the root system. As it is well known, the presence of roots within the soil increases, with respect to the case of soil without vegetation, the material effective cohesion with no significant change in its friction angle. Such mechanical effect can be introduced in the Mohr–Coulomb failure law through an "apparent cohesion" term, which adds to the soil effective cohesion.

The contribution of root reinforcement to the soil shear strength has been investigated in slope stability finite element analysis, modifying the soil properties of individual slope elements including vegetation. This approach allowed quantifying the effect of the mechanical root reinforcement on the slope factor of safety and assessing the sensitivity of slope stability to the variation of apparent cohesion and root zone depth assumed in the numerical simulations.

Keywords: slope stability, root reinforcement, apparent cohesion, finite element method.

1 Introduction

The presence of a root system into the soil plays an important role on the stability of natural and artificial slopes, which are covered with vegetation. It affects the stability of a slope essentially through hydrological and mechanical effects. Regarding the latter aspect, the density of roots within the soil mass and their tensile strength contribute to improve the capacity of the soil to resist shear loads (Figure 1). The maximum tensile strength or pull-out resistance of the roots, together with an assessment of the root size and distribution (Root Area Ratio), can be used to evaluate the appropriate root reinforcement values to be used in the stability analysis of a slope. Many authors have provided values of root systems depth and tensile strength of different species of herbaceous and shrub type. In particular, the experimental data obtained from direct shear tests performed on blocks of soil containing roots have shown that the presence of vegetation produces an increase in soil cohesion, leaving its friction angle unchanged (Wu et al. [1], Faisal and Normaniza [2]).

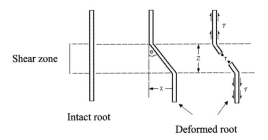

Figure 1: Schematization of root reinforcement (from Gentile et al. [3]).

Such mechanical effect can be introduced in the Mohr–Coulomb failure law through an "apparent cohesion" term, which adds to the soil effective cohesion (Gentile et al. [3]):

$$\tau = (c'+c_R) + \sigma' \tan\varphi' = (c'+c_R) + (\sigma-u)\tan\phi' \qquad (1)$$

where τ is the shear stress on the failure plane, σ' represents the effective stress normal to this plane (equal to the difference between the total normal stress σ and the pore water pressure u), c' is the effective cohesion, c_R the apparent cohesion and ϕ' is the effective friction angle of the soil.

The apparent cohesion c_R can be expressed as (Norris and Greenwood [4]):

$$c_R = 1.2\, T_R\, (A_R/A) \qquad (2)$$

where T_R is the mean tensile strength of the roots and A_R/A is the cross-section of soil occupied by the roots (Root Area Ratio).

Wu et al. [5] have studied the stability of slopes before and after the removal of forest cover, incorporating the apparent cohesion due to roots in the limit equilibrium analysis of infinite slopes. The authors have shown how it is possible in this way to increase the safety factor of the analysed slopes, therefore concluding that the contribution to shear strength provided by the root system is crucial in studying the stability of natural slopes.

Recently, Chok et al. [6] have analysed the mechanical effect due to vegetation on the stability of ideal slopes, using a numerical code based on the finite element method (Zienkiewicz and Taylor [7]). The method, widely employed for the numerical solution of different engineering problems, allows the user to easily define the extent of the vegetation effects, being the slope geometry discretised into small elements. Moreover, the approach provides information about the overall stability of the slope as the value of the factor of safety (FOS) on the critical slip surface can be derived through the $c'-\phi'$ reduction technique (Griffiths and Lane [8]).

The work describes the analysis of the mechanical effect of root systems on slope stability using the finite element code *PLAXIS 2D* [9] and adopting an approach similar to that proposed by Chok *et al.* [6].

2 Tensile and shear strength of roots

2.1 Tensile strength

According to the literature, the values of roots tensile strength generally depend on various factors: species, dimensions, morphology and spatial directions (Figure 2).

Figure 2: Morphological differences between root systems of different shrubby species (from Mattia *et al.* [10]).

Stress–strain curves obtained by traction tests have been processed to obtain the peak tensile strength values. The laboratory data show that the tensile strength generally decreases with root diameter, as reported in Figure 3: root strengths are lower for large diameters and higher for small diameters (Bischetti *et al.* [11], Gray and Barker [12]). Moreover, root strength depends on the biological components of the root: smaller diameter roots have more cellulose than larger diameter roots and therefore are characterised by higher strength (Genet *et al.* [13]).

Regarding the distribution of roots in soil deposits, the observed values of Root Area Ratio (RAR) show a very high variability with species, location and depth. RAR is strongly influenced by genetics, local soil, climate characteristics and forest management; in addition, randomness must be accounted for. However, RAR usually decreases with depth as a consequence of a decrease of nutrients and aeration, and because of the presence of more compacted layers (Bischetti *et al.* [11]).

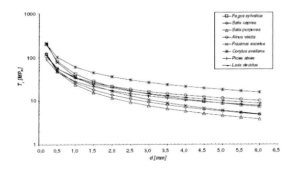

Figure 3: Strength–diameter fitting curves for different species (from Bischetti et al. [11]).

2.2 Apparent cohesion

The reliable benefit of apparent cohesion is limited to shallow depths as root distribution is mainly concentrated within 1m from the ground surface (Figure 4). The use of an enhanced value of the soil cohesion is appropriate for grass and shrub areas where fine root distribution with depth is consistent and easily defined (Norris and Greenwood [4]).

Field studies of forested slopes (O'Loughlin [14]) indicate that the fine roots, 1 to 20mm in diameter, are the ones that contribute most to soil reinforcement. Grasses, legumes and small shrubs can have a significant reinforcing effect down to depths from 0.75 to 1.5m (Faisal and Normaniza [2]).

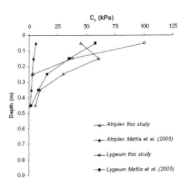

Figure 4: Value of root cohesion at different depths (from De Baets et al. [15]).

Some researchers have attempted to compute the values of apparent cohesion due to the presence of the roots in the ground by designing and developing in situ shear tests for different types of root systems (O'Loughlin and Ziemer [16], Norris and Greenwood [17], van Beek et al. [18]).

3 Vegetation effect on the stability of an ideal slope

The contribution of root reinforcement to soil shear strength has been investigated by numerical stability analyses of ideal slopes in plane strain conditions using the two-dimensional finite element code *PLAXIS 2D* [9]. This approach has allowed to quantify the effect of the mechanical root reinforcement on the slope factor of safety (FOS), assessing the sensitivity of slope stability to the variation of apparent cohesion (c_R) and root zone depth (h_R) assumed in the numerical simulations. In particular, a slope characterised by the absence of water has been initially considered. The presence of vegetation has been simulated by attributing to the elements of the mesh representing the layer with presence of roots a value of effective cohesion higher than the one of the surrounding soil. The weight of the plants has been neglected, as in the case of herbaceous or shrubby vegetation; in the case of trees, it should be taken into account. Subsequently, an ideal slope with a water table at ground surface has been analysed.

3.1 Case A: slope characterised by the absence of water and $c'=0$kPa

The first case studied (Case A) is relative to an ideal slope with an inclination angle β equal to 26.5°, composed by a homogeneous material ($c'=0$kPa, $\phi'=25°$ and $\gamma=20$kN/m^3) and characterised by the absence of water. The geometrical configuration of the slope and the adopted finite element mesh are shown in Figure 5.

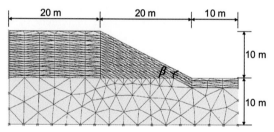

Figure 5:   Adopted mesh for Cases A and B.

At first, the stability of a homogeneous slope without vegetation ($c_R=0$kPa) has been assessed. The results of this analysis have been taken as reference for the evaluation of the mechanical effects due to the presence of vegetation on the slope. Figure 6 shows the contour lines of shear strains at failure obtained at the end of the $c'-\phi'$ reduction analysis. The obtained shear strain values relate to a condition close to collapse and, therefore, have no physical meaning. Nevertheless, they indicate the development of a planar and shallow failure mechanism inside the slope, with a maximum concentration of shear strains at its toe. The depth of the critical surface, measured at the centre of the slope, is equal to 1.4m from ground level. The slope is characterised by a FOS close to one, consistently with what has been obtained by a limit equilibrium analysis.

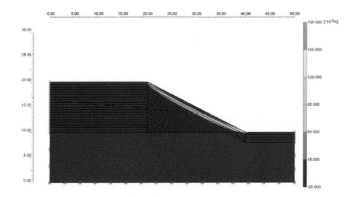

Figure 6: Contour lines of shear strains at failure for a slope without vegetation (Case A).

The effect of the presence of vegetation on slope stability has been initially analysed attributing a value of apparent cohesion equal to 5kPa to a layer of soil elements disposed along the slope surface for a depth $h_R=1m$.

The introduction of vegetation confined along the slope surface only results in a small increment of the safety factor. In particular, the obtained FOS is equal to 1.02, with an increase of 5.9% with respect to the case of slope without vegetation. From the contours lines of shear strains shown in Figure 7, it can be clearly observed how the presence of vegetation produces a downward shift of the critical surface, which is characterized by a depth of 2m. If the slope toe elements are also treated as vegetated soil, the increment of the slope safety factor is more significant. The FOS reaches a value of 1.05 with an increase of 9.2%, while the critical surface is characterized by a depth of 2.4m.

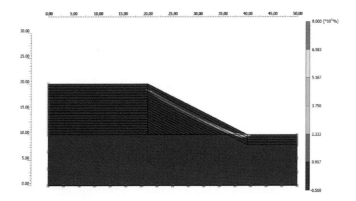

Figure 7: Contour lines of shear strains at failure for a slope with vegetation confined along the slope surface only (Case A).

Finally, the introduction of vegetation extending over the entire ground surface for a depth of 1m gives to the slope a FOS equal to 1.08 (an increase of

11.5% compared to the case without vegetation), producing an additional downward shift of the critical surface to a depth of 2.6m. The analysis shows how the presence of roots distributed uniformly throughout the slope have a positive effect on its stability, with a significant increment of the FOS. The effect increases as much as the root apparatus extends in depth, reaching the zones where the failure mechanism is initiated. Consequently, the critical slip surface is shifted deeper below the ground surface, becoming circular, as shown in Figure 8.

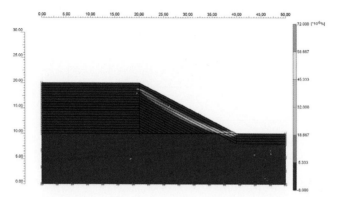

Figure 8: Contour lines of shear strains at failure for a slope with vegetation extending over the entire ground surface (Case A).

Parametric studies have been performed for a range of apparent root cohesion. Figure 9 shows the evolution of the slope FOS with the apparent root cohesion for $h_R=1m$, when $c'=0kPa$ and the vegetation is confined i) to the slope surface only, ii) slope and toe and iii) vegetation extends over the entire ground surface. The values of the critical surface depth with c_R are shown with dashed line in the same figure.

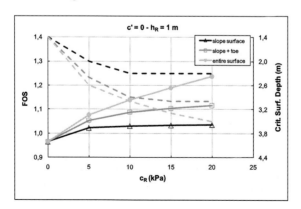

Figure 9: Evolution of FOS and critical surface depth with c_R for $h_R=1m$ (Case A).

Assuming a value of $h_R=2m$, the results of the FE simulations, reported in Figure 10, indicate that the values of FOS and critical surface depth are larger for the same apparent cohesion.

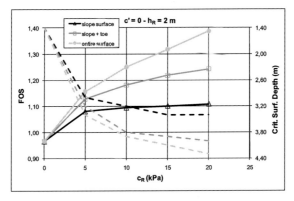

Figure 10: Evolution of FOS and critical surface depth with c_R for $h_R=2m$ (Case A).

3.2 Case B: slope characterised by the absence of water and $c'=5kPa$

Case B regards the same ideal slope of Case A ($\beta=26.5°$, $\phi'=25°$ and $\gamma=20kN/m^3$), but composed of a soil with $c'=5kPa$. In this situation, the limit equilibrium analysis provides a critical slip surface that is no longer parallel to the surface, but circular and deep, with an associated FOS equal to 1.34.

This is confirmed by the contour lines of shear strains obtained through the finite element analysis, as shown in Figure 11. The critical surface starts from the toe of the slope, deepening up to 3.6m from the ground level. In this case, the mechanical effect of vegetation on slope stability has been investigated by assigning a value of apparent cohesion of 10kPa to the soil elements affected by root reinforcement.

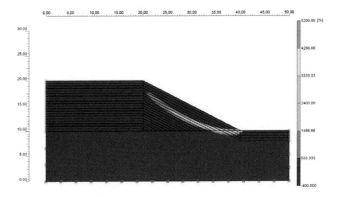

Figure 11: Contour lines of shear strains at failure for a slope without vegetation (Case B).

If the vegetation is only present on the slope surface for a depth $h_R=1m$, the increase in the FOS is just equal to 0.1% with respect to the case of slope without vegetation. The depth of the critical surface does not change significantly (Figure 12). This confirms the limited effect of vegetation on slope stability when the sliding mechanism is deep and the root reinforcement is limited to the first layers below the surface.

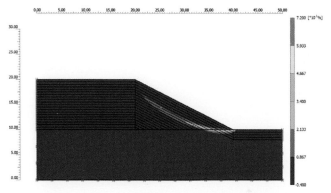

Figure 12: Contour lines of shear strains at failure for a slope with vegetation confined along the slope surface only (Case B).

Assuming the presence of vegetation also at the toe of the slope, the FOS becomes equal to 1.37 (an increase of 2% compared to the case without vegetation) and the failure surface reaches a depth of 3.8m.

Finally, if the vegetation covers the entire ground surface the slope safety factor reaches a value of 1.39, with an increase of 3.2%. The depth of the corresponding failure surface (Figure 13), however, remains almost similar (3.8m) with respect to the case of vegetation covering slope surface and toe.

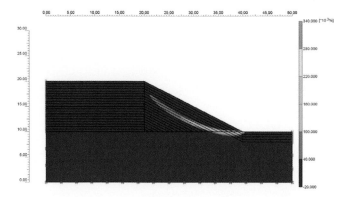

Figure 13: Contour lines of shear strains at failure for a slope with vegetation extending over the entire ground surface (Case B).

Figure 14 shows that, whatever value of c_R is used, the FOS and the depth of the critical surface remain practically unchanged when the vegetation covers the slope surface only (h_R=1m). They increase if the vegetation is introduced at the toe or is distributed over the entire ground surface. The sensitivity analysis indicates that the vegetation mechanical effects are less significant in slopes with high values of effective cohesion where deep-seated failure mechanisms are likely to occur, as the FOS increments with c_R are proportionally lower than the ones obtained in the case of the same slope with c'=0kPa.

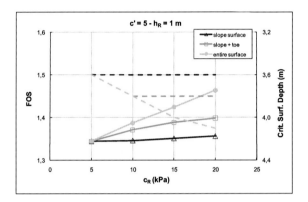

Figure 14:   Evolution of FOS and critical surface depth with c_R for h_R=1m (Case B).

The deepening of the root system to a depth of 2m does not produce any improvement if the vegetation is confined along the slope surface only (Figure 15), unless it extends also to the toe of the slope and to the entire ground surface.

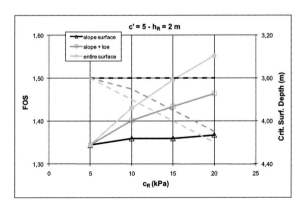

Figure 15:   Evolution of FOS and critical surface depth with c_R for h_R=2m (Case B).

3.3 Case C: slope characterised by the existence of water and $c'=0kPa$

The introduction of a water table at the ground surface in the slope of Case A would produce a significant reduction of soil effective stress, leading to a FOS below one. The slope angle assumed in Cases A and B (26.5°) is, in fact, too high to account for the presence of a water table inside a homogeneous slope characterised by a soil friction angle of 25° and a cohesion equal to zero.

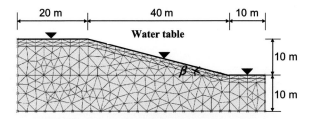

Figure 16:   Adopted mesh for Case C.

The geometry of the ideal slope has been therefore changed, assuming, for the same slope height of 10m, a slope angle of 14° and a soil friction angle of 30°. The adopted finite element mesh is shown in Figure 16.

The reference case is now represented by an ideal slope composed by a homogeneous material with strength parameters $c'=0kPa$ and $\phi'=30°$, characterised by the presence of a water table at the ground surface in steady state conditions and without vegetation ($c_R=0kPa$). In such a case, the FOS of the slope is equal to 1.10 and the corresponding critical surface, which develops from the toe of the slope, is almost parallel to the ground surface, with a depth of about 1.3m (Figure 17).

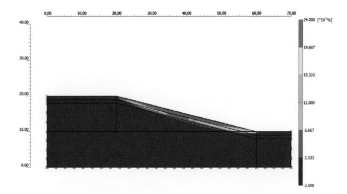

Figure 17:   Contour lines of shear strains at failure for a slope without vegetation (Case C).

As in the previous cases, different distributions of root reinforcement along the slope have been considered, assigning an enhanced value of cohesion ($c_R=5kPa$) to soil elements with presence of vegetation.

When the vegetation covers only the slope surface for a depth of 1m, the FOS slightly increases to 1.13. It corresponds to a failure surface 1.8m deep which remains approximately parallel to the ground surface, as shown in Figure 18. The presence of vegetation produces, therefore, an increase of FOS equal to 2.1% and a deepening of the failure mechanism that is forced to develop below the vegetated soil.

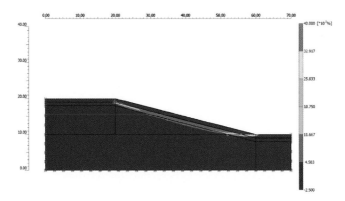

Figure 18: Contour lines of shear strains at failure for a slope with vegetation confined along the slope surface only (Case C).

The effect of the increase of the safety factor and deepening of the critical surface is even more evident if the vegetation includes also the toe of the slope. In this case the FOS becomes 1.16, with an increase of 5.4%, while the depth of the critical mechanism appears to be equal to 2.3m.

The sliding surface deepens (2.5m) when the roots extend over the entire ground surface (Figure 19) and tends to assume a circular shape. The corresponding FOS increases of 7.5% with respect to the case of slope without vegetation, assuming a value of 1.19.

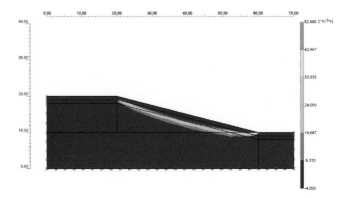

Figure 19: Contour lines of shear strains at failure for a slope with vegetation extending over the entire ground surface (Case C).

As for Case A and B, parametric studies have been performed also in Case C changing the value of the apparent root cohesion. The evolution of FOS and depth of the critical surface with apparent root cohesion (for $h_R=1m$) is shown in Figure 20 and the results are similar to those of Case A. Without vegetation the failure mechanism is shallow and planar. When the vegetation is confined along the slope surface only, the increase of c_R initially produces an increase of FOS and critical surface depth, which remain almost constant afterwards. A continuous increase of FOS can be observed if the vegetation extends over the entire ground surface. Correspondingly, the sliding surface tends to become deeper and circular.

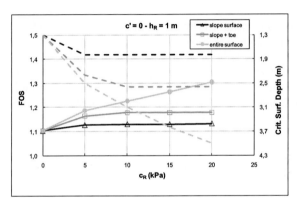

Figure 20: Evolution of FOS and critical surface depth with c_R for $h_R=1m$ (Case C).

The stabilizing effect produced by vegetation is more evident when, for the same apparent cohesion, the depth of root zone is assumed equal to 2m, as shown in Figure 21.

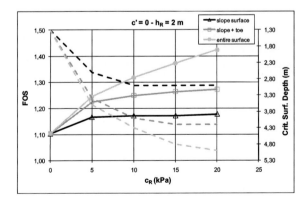

Figure 21: Evolution of FOS and critical surface depth with c_R for $h_R=2m$ (Case C).

4 Conclusions

The work investigates the influence of vegetation on slope stability, with particular emphasis on the mechanical effects due to the presence of roots into the soil. The increase of the slope safety factor provided by root reinforcement has been evaluated using a two-dimensional finite element code enhancing the effective cohesion of individual slope elements with presence of vegetation.

When the failure mechanism inside a slope without vegetation starts from its toe and is planar and shallow, the introduction of vegetation confined along the slope surface only results in small increments of the safety factor. If the slope toe elements are treated as vegetated soil or the vegetation extends over the entire ground surface, the increment of the slope safety factor is more significant. In these cases, the effect increases as much as the root system extends in depth, reaching the zones where the failure mechanism is initiated. Consequently, the critical slip surface is shifted deeper below the ground surface, becoming circular. The sensitivity analysis indicates that the vegetation mechanical effects are less significant in slopes with high values of effective cohesion where deep-seated failure mechanisms are likely to occur. Moreover, the existence of a water table at the ground surface does not generate any considerable change in the general framework observed during finite element analyses of slopes without water.

References

[1] Wu, T.H., Beal, P.E. & Lan, C., In situ shear test of soil-root system. *Journal of Geotechnical Engineering*, **114**, pp. 1351-1375, 1988.
[2] Faisal, H.A. & Normaniza, O., Shear strength of soil containing vegetation roots. *Soils and Foundations*, **48(4)**, pp. 587-596, 2008.
[3] Gentile, F., Romano, G. & Trisorio-Liuzzi, G., L'uso della vegetazione negli interventi di difesa del suolo in ambiente mediterraneo. *Genio Rurale*, **2**, pp. 42-51, 1998.
[4] Norris, J.E. & Greenwood, J.R., Assessing the role of vegetation on soil slopes in urban areas. *Proc. 10th Congress of the International Association for Engineering Geology and the Environment (IAEG)*, Nottingham, UK, 2006.
[5] Wu, T.H., McKinnell, W.P. & Swanston, D.N., Strength of tree root and landslides on Prince of Wales Island, Alaska. *Canadian Geotechnical Journal*, **16**, pp. 19-33, 1979.
[6] Chok, Y.H., Kaggwa, W.S., Jaksa, M.B. & Griffiths, D.V., Modelling the effect of vegetation on stability of slopes. *Proc. 9th Australia New Zealand Conference on Geomechanics*, Auckland, 2004.
[7] Zienkiewicz, O.C. & Taylor, R.L., *The Finite Element Method*, Wiley, John & Sons, 1991.
[8] Griffiths, D.V. & Lane, P.A., Slope stability analysis by finite elements. *Géotechnique*, **49(3)**, pp. 387-403, 1999.
[9] *PLAXIS 2D, Reference Manual*, Version 8, 2003.

[10] Mattia, C., Bischetti, G.B. & Gentile, F., Biotechnical characteristics of root systems of typical Mediterranean species. *Plant and Soil*, **278**, pp. 23-32, 2005.

[11] Bischetti, G.B., Chiaradia, E.A., Simonato, T., Speziali, B., Vitali, B., Vullo, P. & Zocco, A., Root strength and root area ratio of forest species in Lombardy (Northern Italy). *Plant and Soil*, **278**, pp. 11-22, 2005.

[12] Gray, D.H. & Barker, D., Root-Soil Mechanics and Interactions, *Riparian vegetation and fluvial geomorphology*, Water Science and Application 8, American Geophysical Union, Washington DC, pp. 125-139, 2004.

[13] Genet, M., Stokes, A., Salin, F., Mickovski, S.B., Fourcaud, T., Dumail, J.F. & van Beek, R., The influence of cellulose content on tensile strength in tree roots. *Plant and Soil*, **278**, pp. 1-9, 2005.

[14] O'Loughlin, C.L., Effectiveness of introduced forest vegetation for protecting against landslides and erosion in New Zealand's steeplands. *Proc. Symposium on effects of forest land use on erosion and slope stability*, Honolulu, Hawaii, 1984.

[15] De Baets, S., Poesen, J., Reubens, B., Wemans, K., De Baerdemaeker, J. & Muys, B., Root tensile strength and root distribution of typical Mediterranean plant species and their contribution to soil shear strength. *Plant and Soil*, **305**, pp. 207-226, 2008.

[16] O'Loughlin, C.L. & Ziemer, R.R., The importance of root strength and deterioration rates upon edaphic stability in steepland forests. *Proc. I.U.F.R.O. Workshop P.1.07-00 Ecology of subalpine ecosystems as a key to management*, Oregon, USA, pp. 70-78, 1982.

[17] Norris, J.E. & Greenwood, J.R., In-situ shear box and root pull-out apparatus for measuring the reinforcing effects of vegetation. *Proc. Field Measurements in Geomechanics*, Swets and Zeitlinger, Lisse, pp. 593-597, 2003.

[18] van Beek, L.P.H., Wint, J., Cammeraat, L.H. & Edwards, J.P, Observation and simulation of root reinforcement on abandoned Mediterranean slopes. *Plant and Soil*, **278**, pp. 55-74, 2005.

Experience with treatment of road structure landslides by innovative methods of deep drainage

O. Mrvík[1] & S. Bomont[2]
[1]*Department of Geotechnics, Czech Technical University in Prague, Czech Republic*
[2]*TP.GEO, France*

Abstract

During construction of infrastructure projects, emergency situations due to the presence of groundwater and slope instability occur very often. In addition, damage to existing roads or railways caused by groundwater or slope deformations do not represent any extraordinary situation. In such cases, the water should be taken from the ground in order to improve the properties of the soils and rocks. According to the consumption of energy, the methods of dewatering can be divided into gravity dewatering and dewatering with electricity. Traditionally used drainage techniques are proven methods. However, in certain geotechnical conditions, two innovative drainage systems, Siphon Drains and Electropneumatic Drains, can prove to have many advantages. In this paper, several applications of the innovative deep drainage systems are described. The paper introduces selected sites where groundwater lowering played a key role in the successful solution of slope stabilization and road remediation. The sites are located in France.

Keywords: road structures, landslide, groundwater, deep drainage, Siphon Drains, Electropneumatic Drains.

1 Introduction

Road structures, as roads, highways or railways, are very often endangered or even seriously damaged by different kinds of deformations of natural slopes or artificial excavations. The main trigger for such deformations lies in high groundwater levels within the affected area or just the subsoil of the road embankment. Therefore, groundwater lowering is primarily the most important measure in order to avoid the risk of slope movements, damage to the road structures and, secondarily, to protect the affected areas against repeated deformation and destruction.

Geotechnical designers responsible for the treatment of such problems should always recognize the main reasons for the situation, evaluate the geotechnical conditions of the site and look for the most effective solution. The decision-making process of a proper drainage system to drain and stabilize the affected site might be as follows.

Traditional methods of dewatering, such as drainage trenches, are usually limited by their maximal economical depth of 3–5 m. Deeper excavations would cause huge and non-economical earthworks and extra expenses for gravel backfilling. In the case of subhorizontal wells, considerable length of drilling, difficulties in reaching all aquifers and problems of site access can be considered as significant disadvantages. Alternatively, the innovative method of Siphon Drains®, which allows dewatering up to 8–12 m beneath the surface without the need for electrical energy, can be adopted.

Deep drainage and groundwater lowering deeper than 10 meters today still represents a great deal of effort and implies a difficult technique. However, sometimes it is inevitable to reach deeper aquifers that might cause many serious problems in construction and to deal with groundwater lowering of tens of meters. Traditional techniques are very often badly fitted to achieve such requirements. Submersible pumps require a rather high minimum flow and frequent expensive maintenance. Well points efficiency is limited to a depth of 7 meters lower than the level of the vacuum pump. The innovative method of Electropneumatic Drains® has been developed to lower the water table up to 40–60 meters below ground level using pneumatic energy (compressed air).

The groundwater problem at the first three of four introduced sites was treated by Siphon Drains as a permanent energy-free solution. The last case study introduces an application of Electropneumatic Drainage as a permanent solution to an emergency problem that occurred soon after construction of a project of highest importance.

2 Innovative methods of deep drainage

2.1 Siphon drains

Small diameter (10–30 mm) suction siphon tubes are placed into vertical wells. The wells are spaced between 3–6 m and are sufficiently deep to provide required drawdown. The wells are dewatered using the siphon principle based on gravity drawdown up to depths of 8–12 m beneath the surface (Fig. 1).

The tubes are inserted into a permanent water filled reservoir at the base of each well with an outlet downstream at an outlet manhole, situated down slope. If the water level rises in the well, the flow in the tube is renewed and abstracts water out of the well. The flow continues until the water level in the well falls back to the reference level, providing that the flow rate in the siphon is sufficient to keep the siphon primed. As the water rises towards the top of the tube, the

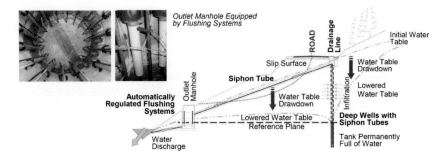

Figure 1: Basic principles of Siphon Drainage.

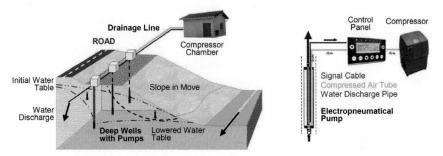

Figure 2: Arrangement of Electropneumatic Drainage.

pressure falls and may reach vacuum, causing the creation of bubbles. Without any additional measure, the bubbles might cause the flow to break. This is avoided by using the flushing system, which flushes bubbles out by turbulent flow and controls and regulates the flow so that the siphon always stays primed. The flushing system is an arrangement of PVC pipes and it is placed downstream in the outlet manhole at the end of each siphon tube.

The method is suitable for the geological environment of permeabilities less than 10^{-5} m/s and groundwater inflows of 0.0–2.0–15.0 l/min per well (i.e. 0.0–0.03–0.3 l/s per well). The main advantages are: the system is automatically continually in service, it is easily controlled, of high efficiency and can facilitate drawdowns of up to 12 m, without the need for any additional energy.

2.2 Electropneumatic Drains

The drainage is designed by a network of vertical wells, manholes, ducting for cabling and pipes for water discharge and compressed air (Fig. 2). The wells are spaced between 3–6 m and are sufficiently deep to provide the required drawdown. The wells are equipped with a casing of 110 mm diameter and a gravel filter. Electropneumatic pumps are installed into the wells at a defined depth, connected to a compressor and the control panel and equipped with an intelligent sensor. When the groundwater rises in the wells, it fills the pumps and when it reaches the high level sensor, an electrical signal is transmitted to the control panel, including the relays and solenoids. The signal triggers the injection of compressed air into the pumping chamber to push water out onto the surface through an outlet tube. The pump filling and emptying is repeated until the required drawdown is reached.

The system manages groundwater lowering of up to 60 m under permeabilities of 1.10^{-5}–1.10^{-7} m/s with groundwater inflows of 0.0–35.0 l/min per well (i.e. 0.0–0.6 l/s per well). The main advantages are: the system runs only under high groundwater level, the pumps contain no moving mechanical parts and both operation and maintenance expenses are saved.

3   RD74 – Goncourt

At this site, the affected section of the road is situated on a slope above a shallow valley with a river at the bottom (Fig. 3). The valley is filled mainly by soft sediments, such as clays. The bedrock is created by alternating marls and limestones. At the top of the slope, the limestones are exposed either by natural outcrops and cuttings made during construction of the road. The limestones are affected by karstic effects and by mechanical weathering. A complicated system of the groundwater of many aquifers confined mainly in cracks and voids is developed within the limestone formation. Infiltration of surface water was allowed. The water flow follows the slope so that the soft sediments in the lower parts of the valley were saturated. The water that occurred in the clays was very shallow, 0–5 m below surface. As the slope is being undercut by the river erosion, deformations in the saturated clays were initiated easily. These deformations were accompanied by gravity moves of limestone blocks in the upper part of the slope. Downstream movements of the limestone blocks caused deformations of the road. The affected area was 300 m in width and 150 m in length. The deepest deformations were expected to be 10 m.

To eliminate the deformations and to avoid future road destructions, the water coming from the top of the slope had to be drained so that the shear properties of the clays in which sliding occurred would be improved. Predominantly, long-term permanent gravity drainage without the need for energy was required. Pumping of the water from deep wells situated at the top of the slope would have been possible, but very time consuming. Continual water flow of a huge amount was expected. Horizontal wells carried out from the bottom of the slope would have to be very long. Reaching all of the required aquifers would not be guaranteed by horizontal boreholes. A drainage trench situated below the top of the hill was taken into account. Such a trench would have to be very deep to attain the deepest aquifer. The earthworks and the gravel backfill of such an excavation would be extremely expensive. Moreover, it might be dangerous to cut the slope by a 5–10 m deep trench that could break the weak stability or suddenly bring a lot of groundwater from cracks to the slope. The system of Siphon Drains was selected as the most suitable and effective.

A single drainage line of 71 drains was situated in the slope as an artificial water barrier (Fig. 4). The purpose of the drainage was to take the incoming water out of the slope and to lower the water table permanently. A shallow trench of 1.2 m minimum depth in which to place all necessary ducting, as well

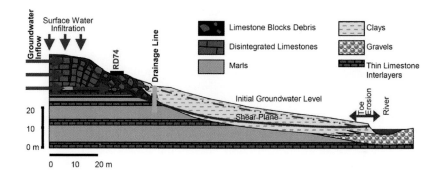

Figure 3: Cross-section of the affected slope at Goncourt.

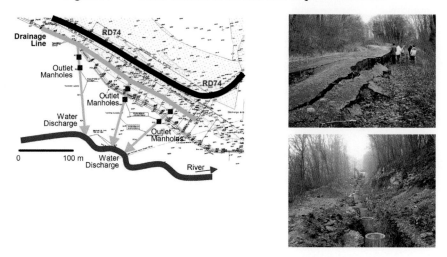

Figure 4: Scheme of the Siphon Drainage System at Goncourt.

as to protect the equipment of the Siphon Drainage technology against frost action and mechanical destruction, was excavated first. Protective drainage manholes made of concrete prefabricates Ø800 mm were placed at the position of each drain. The trench was partially backfilled by gravel and a perforated plastic duct was placed at the bottom to be used as a surface drain. The wells were drilled at Ø250 mm from the top of the manholes after backfilling the trench. The drains are 12.5 m deep and spaced at 5 m. The wells were equipped as standard opened piezometers. A perforated plastic pipe Ø110/114 mm was inserted into the boreholes and the space between the borehole and the screen was filled with filtrating gravel of 2–4 mm. An Air-lift to clean the wells was applied.

The collected groundwater was directed through siphon pipes (one for each drain) in three evacuation lines. The water collected by the drains is sucked out of the wells by the natural siphon principle through 10/12 mm diameter plastic siphon pipes inserted into each well and pulled through the burned ducting pipes between neighbouring manholes to three crosspoint manholes and then to six downstream placed common outlet manholes. Flushing systems for each drain were installed in the outlet manholes. The siphon tubes were primed by water under pressure and the outlet endings of the siphon pipes were connected to relevant flushing systems. A system of continual groundwater monitoring by piezometers was established.

After the system was put in function, the groundwater level in each well was lowered to 8–11 m below the terrain. In the initial stages of the drainage function, the global water discharge reached 250 l/min (3.5 l/min per drain). The current overall values of the water flow through the system are around 70 l/min (1 l/min per well). As of 2010, the system had already been in operation for seven years without any defects. Regular periodical maintenance of the drainage is carried out. No significant problems with the stability of the area occur anymore.

4 RD95 – Aigueblanche

On the national road RD95, for 3.680–3.800 km, considerable ground movements and problems with instability have been occurring for more than ten years. The instability was characterized by several active landslide zones developed on a steep slope in the altitudes between 700–850 m above sea level (Fig. 5). The rate of deformations varied from 10 to 70 mm per year and according to the inclinometry measurements, the depth of active shear planes varied between 6 and 14 m beneath the surface. The geological settings were designated as favourable for groundwater circulation and surface water infiltration. The ground is created by mixed rockmass and soils and it is composed of shales of variable degrees of weathering and clays. The initial piezometric levels were observed at a few levels between 1.5–3.0 m.

The results of a research study were that the sliding is too complex to stabilize the slope entirely by implementing mechanical barriers as pile walls or "Berlin" walls or some other method. The water table had been observed to be too high at the site. Geotechnical studies considered that the establishment of a drainage network with an 8 m depth efficiency should allow certain improvement of the safety factor, but still the drainage solution would not allow the complete stabilization of the slide. However, this solution was expected to slow down the movements and thereby reduce the deformations of the roadway. The gravity driven drainage system by means of Siphon Drains was chosen to achieve these goals.

Since the purpose of the drainage was just to reach a certain groundwater drawdown and the scale of the affected area was too large, exceptionally, the Siphon Drainage line was placed in the middle of the instable slope, below the road (Fig. 6). The drainage network was created by a single line of a total length of 130 m by 26 vertical Siphon Drains equipped with protective manholes (800 mm in diameter) with a spacing of 4 m. The drainage trench (for construction of the manholes, placing a collector at Ø200 mm for siphon tubes and a perforated collector at Ø150 mm for surface water) was up to 2.5 m deep (to realize the

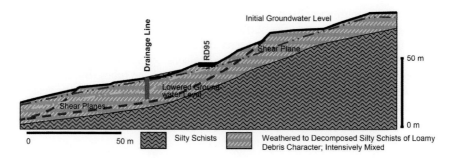

Figure 5: Cross-section of the affected slope at Aigueblanche.

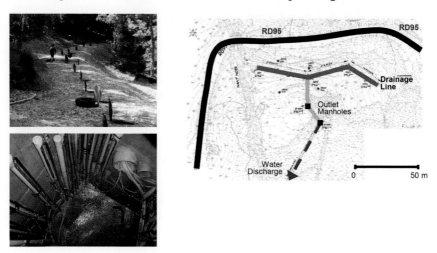

Figure 6: Scheme of the Siphon Drainage System at Aigueblanche.

drawdown as deep down as possible). After construction and backfilling of the trench, the boreholes were drilled up to 15 m beneath the surface. Two outlet manholes were executed in the slope below the drainage line. The outlet manholes were constructed from prefabricated rectangle concrete units of a 1500 mm. The outlet manholes were connected to the drainage line by the same type of trench as the trench between the drains, including a Ø200 mm ducting pipe (collector) for siphon tubes and a Ø150 mm perforated collector for surface water. The drains were constructed and equipped as standard opened piezometers. The diameter of drilling was 250 mm. A perforated plastic pipe Ø110/114 mm was inserted into the boreholes and the space between the borehole wall and the screen was filled with gravel of grading 2–4 mm. The water collected by the drains is sucked out of the wells by the natural siphon principle through 10/12 mm diameter plastic siphon pipes inserted into each well. After the lower situated outlet manhole, the collected ground and surface water was directed by gravity downstream, to an opened concrete ditch. A system of continual groundwater monitoring by piezometers was established.

After the system was put in function, the groundwater level in each well was lowered to 10–12 m below the surface. The maximal registered flow for one single drain was 1.7 l/min. For reasons of extreme water inflows into some drains, the capacity was increased at these places by additional siphon tubes inserted into the wells or by changing the diameter of the tubes. By doing this, the maximal managed water flow was increased to 15 l/min. In 2010, the system had been in operation for four years without any defects. Regular periodical maintenance of the drainage is carried out. No problems with stability of the area occur anymore.

5 RD104 – Saint Priest

The road in the vicinity of Saint Priest was deformed due to general slope instability caused by a high water table (Fig. 7). A section of approximately 100 m of the state road RD104 was in danger if nothing was done.

The bedrock at the site is characterized by marls of different degrees of alteration and different degrees of compaction. The marls are alternating with limestones. It is though that the rockmass is slightly jointed. Quaternary deposits are developed as heterogenous debris, the thickness of which increases uphill from 0 m below the road embankment to 5 m in the slope above the road. This was caused by earthworks during the road construction. The upper groundwater level should have been lowered and was initially following the boundary between the debris and the marls. It was also observed to be very shallow in the marls below the road embankment. The upper water level was oscillating strongly according to precipitation. The lower groundwater level was explored in the marl formations, but it was thought not to have an influence on the stability problems.

In order to guarantee the water table lowering, the Siphon Drains method was selected as the most convenient for permanent water drawdown without electricity consumption. By using vertical wells, the water could have been easily lowered up to the depths required by the geotechnical designer in order to improve the general stability of the slope (Fig. 8).

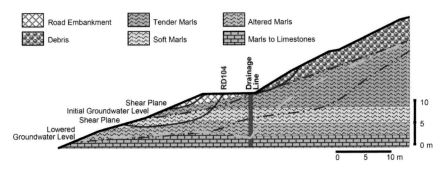

Figure 7:　　Cross-section of the affected slope at Saint Priest.

The drainage network was arranged in a single line containing 37 drains. The line was placed just next to the road. The standard arrangement of the Siphon Drainage system was kept at this site. A trench of 1.5 m minimum depth in which to place all necessary ducting, as well as to protect the equipment of the Siphon Drainage technology against frost action and mechanical destruction, was excavated. To realize the drawdown as deep down as possible, the trench was excavated up to 2 m depth. Protective drainage manholes made of concrete prefabricates Ø800 mm were placed at the position of each drain. The trench was completely backfilled with gravel. A perforated plastic duct was placed at the bottom to be used as a surface drain. The wells were drilled at Ø250 mm from the top of the manholes after backfilling the trench. The drains are 13.5 m deep. The distances between the drains are 3 m. The wells were equipped as standard opened piezometers. A perforated plastic pipe of Ø110/114 mm was inserted into the boreholes and the space between the borehole and the screen was filled with filtrating gravel of 2–4 mm. An air-lift to clean the wells was applied.

The collected groundwater was directed through siphon pipes (one for each drain) in two evacuation lines. The water collected by the drains is sucked out of the wells by the natural siphon principle through 10/12 mm diameter plastic siphon pipes inserted into each well and pulled through the burned ducting pipes between neighbouring manholes to two crosspoint manholes and then to four downstream placed common outlet manholes. Flushing systems for each drain were installed in the outlet manholes. The siphon tubes were primed by water

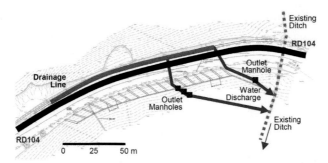

Figure 8:　　Scheme of the Siphon Drainage System at Saint Priest.

Figure 9:　　Deformations of the road and view at the Saint Priest site.

under pressure and the outlet endings of the siphon pipes were connected to relevant flushing systems. The final water evacuation was realized to an existent opened ditch.

A system of continual groundwater monitoring was established by means of 14 vibrating wire piezometers placed into the same wells. Telemetry was set up so the data could be downloaded, checked and processed anytime in the office and provided online to the client. After the system was put in function, the groundwater level in each well was lowered to 9–13 m below the surface. At the moment of preparing this article (soon after installation of the system), no information about water flows were known. Regular periodical maintenance of the drainage is carried out.

6 TGV Corridor – Chabrillan

The "Chabrillan" is a major cutting excavated in 1998 on the TGV high speed train link in France. It is located 530.300,20 km south of Valence. The cut is 1000 m long, with a maximum depth 35 m. In summer 2000, minor deformations on the access road were noticed. Later on, larger scale movements of a few centimetres in width resulted in a narrow fissure on the slope above the road. By the end of 2000, the fissure developed into a major feature of 30 m in length. It was suggested that movements and slope failures could affect the cutting and the train corridor. In 2001, the maximum recorded lateral ground movement reached about 1 mm/day towards the rail line and a total volume of 1.200 000 m^3 was in movement along two shear planes that were identified at the site.

The geology comprises molasse deposits principally formed by sandstones and fresh water limestones with the presence of karstic features and interlayers of plastic clayey marls (Fig. 10). Alpine tectonics is apparent by the presence of thrust planes inclined at 15–20° towards the cutting. Groundwater was observed at two levels, generally at depths of 20 m and 33 m below the crest of the cutting. Inclinometer records confirmed the presence of two main levels of ground movement at 19 m and 30 m. The total volume of the material in movement was indicated as 1.200.000 m^3. It was proposed to make an unloading cut and to excavate 600.000 m^3 of material in the slope behind the railway cutting to relieve the disturbing forces driving the slope instability. After performance of the excavation, the groundwater level was monitored in 2001–2005 and observed at 0–8 m beneath the base of the cut. In 2002, new movements were registered.

To control the groundwater level and to reduce the deformations, several dewatering schemes were considered. A 10 m deep trench for a length of 150 m was designed to be excavated from the base of the unloading cut. This solution suffered from a number of limitations: in particular, the trench might increase the risk of new shear failures. Subhorizontal wells were rejected due to considerable length and minimal efficiency due to generally low permeability and complex aquifers. To achieve the required drawdown, a deep drainage system was assumed as the most appropriate approach. Immerged pumps were not selected due to their poor efficiency at low permeabilities and inflows and for high energy requirements. The limitations of the gravity Siphon Drainage system lay in the

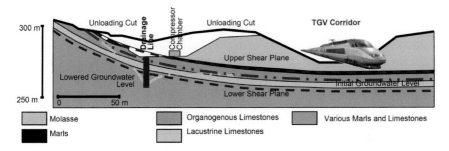

Figure 10: Cross-section of the affected slope at Chabrillan.

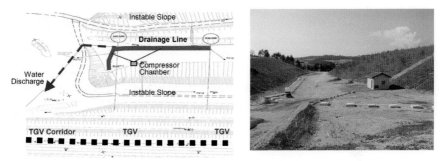

Figure 11: Scheme of the Electropneumatic Drainage System at Chabrillan.

plain gradient of the ground surface. Based on the slide characteristics, the site morphology, the required drawdown and the total construction, operation and service expenses of the drainage, a scheme by Electropneumatic Drains was selected as the most effective solution.

The design comprised 47 wells arranged in a 150 m long line (Fig. 11). The wells were bored at 200 mm diameter into 12–20 m depth and equipped with electropneumatic pumps. A slotted PVC casing of 103 mm internal diameter was installed to full depth. The annulus was filled with a fine graded gravel filter of 2–4 mm size. Each drain was equipped with a protective manhole of 1.5 m depth. All tubings and cables were led from wells to the compressor chamber where the compressors (2x30 kW, one as a back–up), the control panel and the accessories were installed. The chamber was designed as a simple brick house (4x6 m). A comprehensive system of instrumentation and monitoring was established. The monitoring comprised continual water level measuring, inclinometry and water flow observations. An alarms system by GSM was set-up for alerts in the case of high water level or any disconnection problems. Internet Explorer was chosen as the interface for the drainage system operation and monitoring results online checking.

The groundwater was lowered from the original 0–8 m to 11–15 m beneath the base of the unloading excavation. The drawdown and drainage efficiency comply with the requirements. The maximal total flow rates recorded reached 8–280 l/min (0.2–6 l/min per drain). The actual moderate flow rate alternates at around 0.2 l/min per well (8 l/min in total). In 2010, the system had been in operation for four years without any defects. Regular periodical maintenance ofthe drainage is carried out. No new slope deformations were registered after start-up of the system. In 2010, a special French price "Ivor" for quality and for innovative solution was given to Electropneumatic Drainage at this site.

7 Conclusions

Corridors of roads, highways or railways are frequently surrounded by slopes – natural hills or artificial cuts. The presence of groundwater in the slopes is one of the most negative factors that can affect the stability and functionality of the infrastructure tracks. Groundwater lowering is one of the ways to stabilize slope movements by improving the properties of soils and rocks and avoiding or reducing the risk of new deformations and damages of the structures. The method of groundwater lowering is very important. In this article, two innovative alternatives of dewatering by systems of deep vertical wells were introduced.

Since many of the stability problems of roads appear in non-urbanized areas without good access for a source of power, the gravity method of Siphon Drainage seems to be an ideal and very effective solution for long-term permanent groundwater lowering without any need for electricity for water pumping. In the case that the water table should be reduced by up to 8–12 m beneath the surface and the expected water inflows are around 2–15 l/min per well, this method provides a reliable solution.

The method of Electropneumatic Drains represents a flexible system of dewatering which does not require any specific morphology of the treated sites. It is suitable for urgent, emergency and even temporary solutions of deeper seated problems (up to 60 m depth). In the case that the groundwater level is supposed to be reduced in cycles (continuous pumping is not expected), the expenses for electricity and maintenance of the equipment can be saved by use of a single conventional compressor (10–50 kW/ 10–15 bars for 20–150 wells up to 60 m depth) instead of many submersible pumps of the same rated power input. A huge amount of water per well (up to 70 l/min) can be managed to be pumped.

References

[1] Bomont, S., Mrvík, O., Back Experience from Two Cases of Stabilization of Coastal Landslides by Innovative Deep Drainage Systems, *Proc. of the 11th Baltic Sea Geotechnical Conference*, Vol. 1, pp. 19–26, 2008.

[2] Bomont, S., Mrvík, O., Back Experience of Innovative Deep Drainage Systems for Slopes Dewatering and Landslides Stabilizations, *Geoinzynieria – drogi, mosty, tunele*, 04/2008 (19), pp. 60–65, ISSN 1895–0426, 2008.

[3] Mrvík, O., Bomont, S., Application of Innovative Method of Deep Drainage by Siphon Drains for Stabilization of Slopes of Former Opened–Cast Brown Coal Mine "Most – Lezaky" (Czech Republic), *Geotechnika,* 2/2009, pp. 20–25, ISSN 1211–913X, 2009.

Strategic program for landslide disaster risk reduction: a lesson learned from Central Java, Indonesia

D. Karnawati[1], T. F. Fathani[1], B. Andayani[1], P. W. Burton[2] & I. Sudarno[1]
[1]*Gadjah Mada University, Indonesia*
[2]*University of East Anglia, UK*

Abstract

The Indonesian Archipelago is one dynamic volcanic arc region, where landslides frequently occur during the rainy season. Not only are geological conditions and high precipitation in the region, but also uncontrolled land use development and high social-vulnerability of the community living in landslide prone areas, that have become the major cause of landslide disasters in Indonesia. Accordingly, a strategic program for landslide risk reduction has been carried out by establishing an appropriate landslide risk management program with respect to social vulnerability. Such programs mainly emphasize the improvement of community resilience in landslide prone areas through community based landslide mitigation and early warning system, as well as public education. Geological investigations combined with social survey and analyses were also carried out to support the implementation of this risk reduction program in Central Java. Finally, it was concluded that the effectiveness of landslide disaster risk reduction was mainly driven by the community empowerment for disaster prevention and mitigation at the village level.

Keywords: landslide mitigation, risk reduction, resilience improvement.

1 Geology and landslide conditions of the study area

Java, as illustrated in fig. 1, which is modified from [1–3], is a dynamic volcanic island which is predominantly covered by Tertiary to Quartenary andesitic-volcanic and carbonaceous-marine sediments with quite intensive structural geology such as fold, faults and joint formation, fig. 2. Interaction of these geological conditions with high rain precipitation (up to 100 mm/ hour or 3000 mm per year) brings about a high susceptibility of landslides in this region. Fig. 3 illustrates the distribution of landslide susceptibility in Central Java, mainly analyzed based on the morphological and geological conditions. It is apparent that the zone with landslide susceptibility covers about 60% of the region in Central Java.

Human interference such as through uncontrolled land use development in such a susceptible region seriously increases the frequency of landslides, especially during heavy rainstorms, as stated by Karnawati et al. [4]. Furthermore, Karnawati and Fathani in [5] also reported that some landslides occurred due to earthquakes, such as those induced by 6.3 Mw Yogyakarta Earthquake of May 27, 2006 along the jointed andesitic and tufaceous rock slopes at the Southern Mountain in Regencies of Bantul, Gunung Kidul and Sleman.

Based on geological field investigation, it can be identified that landslides in Central Java mostly occured as the sliding or falling/ rolling of rock mass such as of bedded or jointed rocks, fig. 4, as well as the sliding or flowing of soils and/or mixture of rock and soil masses, fig. 5, with a high rate of movement up to 25 m/ second. These rapid landslide types mostly occur in steep mountains and hillslopes (with 30° to 60° slope inclination), and they result in a lot of fatalities and casualties. Indeed, rain-induced landslide disasters have resulted in 1211 fatalities, hundreds of casualties and a substantial socio-economical loss since the year 2000. On the other hand, slow movement of soil mass, such as soil

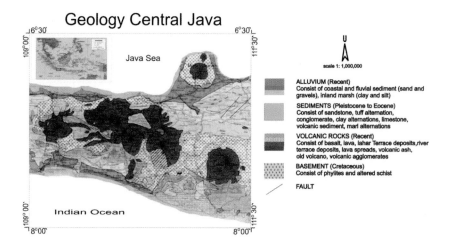

Figure 1: Geological map of Central Java modified from G.S.G.S. [1]; Surono and Sudarno (Surakarta-Giritontro Sheet) [2]; Sampurno and Samodra (Ponorogo Sheet) [3].

Figure 2: Jointed tufaceous sandstone, which is susceptible for rock sliding or falling exposed in Sengir Village, Sleman Regency after the earthquake of May 27, 2006.

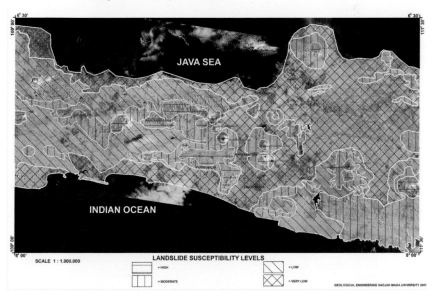

Figure 3: Landslide susceptibility map of Central Java developed based on satellite image interpretation.

creep, has never caused any fatalities but destroyed masses of land and houses at the foot of mountains or hills, fig. 6. This slow movement is mainly induced by the saturation of sensitive clay layers in the slope, which are rich in montmorillonite (smectite and/ or illite).

Figure 4:　　Earthquake induced rock fall in Bantul Regency in May 2006.

Landslides 95

Figure 5: Earth slide (sliding of soil) in Garungang Village, Cimanggu District, Cilacap Regency, occurring in February 2009.

Figure 6: Soil creep which destroyed houses and infrastructure in Tengkllik Village at Karanganyar Regency (top) and in Cimanggu District at Cilacap Regency (bottom) on February 2009.

2 Socio-economical vulnerability

It was apparent that most of the landslide prone areas have been developed with inappropriate land use management. Such areas are normally situated on slopes or at the foot of mountains, covered by thick fertile soils with plenty of water resources and a beautiful panorama. Those conditions attract more and more people to stay and to cultivate the land without having proper landslide awareness. Indeed, more and more landslide prone areas have grown as a densely-populated zone.

Based on the social survey conducted in several selected landslide prone areas, it was found that most of the community living in those areas did not have appropriate knowledge about landslide phenomena nor had immediate access to obtain information for landside mitigation, prevention and early warning. Despite the availability of landslide hazard map established by the national government which provides information about the distribution of landslide susceptibility zone, it is still not yet well-disseminated in an appropriate scale for local mitigation and is too technical, so it was quite hard for the community to understand such information. Indeed, more and more strategic infrastructures have been developed in landslide prone areas with poor consideration for landslide mitigation and early warning systems. Uncontrolled cutting of susceptible slopes for housing, road, and land cultivation is commonly done because of quite limited areas available for landuse development.

Obviously, all the above conditions bring about high socio-economical vulnerability which increases the risk of landslide disasters. Moreover, illegal logging of trees and forest, as well as uncontrolled mining excavation intensify the vulnerability of the landslide prone area.

3 Strategy for landslide disaster risk reduction

It was apparent that Central Java is considered as a high risk area for landslide disaster due to the susceptibility of natural conditions such as geology and climate conditions, as well as due to high vulnerability of the socio-economical conditions in landslide prone areas. Admittedly, it is impossible to change the natural conditions in order to reduce the landslide susceptibility, but it will be more feasible to manage the social conditions for reducing socio-economical vulnerability in landslide prone areas. Therefore, landslide disaster risk reduction in Central Java was conducted by adaptive management as suggested by Andayani *et al.* in [6], which emphasizes vulnerability reduction for community resilience improvment, through both technical and social approaches.

It was also identified that a gap of knowledge and skill between national and local government/communities is one significant obstacle in disaster risk reduction efforts. This gap crucially results in ineffective implementation of disaster risk reduction programs at a local level. Thus, community empowerment and capacity development at the village level was considered to be the most critical driving force for the success of disaster risk reduction programs.

4 Technical-strategic approach

The technical approach was carried out by mainly focusing on providing an immediate access of information related to landslide mitigation and prevention for the local government and community.

It was suggested in [7] that the development of a landslide hazard or the landslide susceptibility map with a medium scale such as 1 : 25,000 illustrated in fig. 7-top is crucial for spatial prediction of landslide susceptibility, in order to facilitate an appropriate development of a landuse management plan at the district and village levels. Accordingly, the development of any infrastructure, houses and services at the zone with high landslide susceptibility should be avoided or minimized, unless, a special engineerng countermeasure is provided. Furthermore, a landslide risk map at the same scale (1:25,000) as illustrated in fig. 7-bottom was also required to estimate the potential socio-economical impacts of any probable landslide occurrence, so that the appropriate landslide risk management can be prepared. However, to ensure that both technical hazard and risk maps can be easily understood and used by the local community for landslide disaster risk reduction, a community landslide hazard and risk map at the local village-scale was required to be developed through community participation. Once the high susceptibility zone can be identified from a landslide hazard map, an appropriate landslide prevention program can be developed. Considering that the coverage area of a landslide susceptible zone was quite large whilst the budget available for landslide prevention was limited, the application of bio-engineering combined with an appropriate drainage system and simple engineering structures was considered to be the most efficient and effective for landslide prevention. It was also recommended that existing indigeneous technology and traditional knowledge should be addressed as well, in order to guarantee the effectiveness of landslide prevention programs.

5 Social-strategic approach through capacity development

5.1 Public education for disaster risk reduction and sustainable development

In order to ensure that landslide disaster risk reduction at the local village can be performed effectively, the village community should be empowered to enable them to participate actively in the disaster management program. Therefore, public education for community empowerment was conducted. It is crucial that such empowerment is not only driven for the disaster risk reduction efforts but also is directed with respect to the sustainable development and livelihood improvement in landslide a prone area.

The public education was specifically designed for the development of practical knowledge about landslide phenomena such as the cause of landslide, how to recognize the susceptible zone and the symptoms of landslide, how to prevent and control landslide occurrence, how to maintain and protect the slope and the village environment so that landslides and other related disasters can be

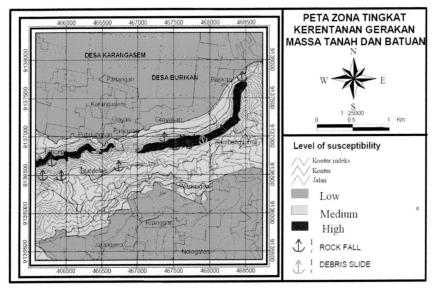

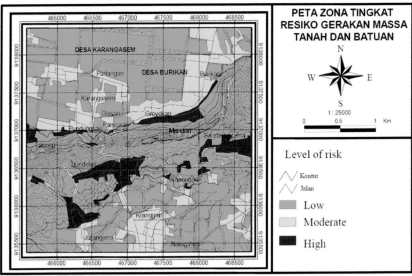

Figure 7: Landslide susceptibility map (top) and risk map (bottom) of Mundon Village Area, Gunung Kidul Regency developed by Mahendra and Karnawati in [7].

avoided. Achievement motivation training was also conducted to increase the community awareness and willingness for successfully conducting landslide prevention program.

5.2 Establishment of community task force for disaster mitigation and management

In order to sustain the community disaster mitigation and management in the landslide prone village, a community task force needs to be established as a key driving force in conducting the village disaster management program [8, 9]. This community task force team may include the head of the village as the team leader, the representatives from local community groups, and the young leaders. This team has an important roles for a continuing public education and community empowerment program, as well as for coordinating the village disaster management program within the village community and with the district/regency disaster task force. One of hardest challenges for ensuring the active performance of this task force is about the continuing spirit and willingness of the community task force in conducting their mission without any financial rewards. A socio-cultural approach such as through traditional and/or spiritual media needs to be addressed to support the sustainability and effectivity of a community task force in conducting the mission for village-disaster management.

6 Socio-technical strategic approach: community-based landslide early warning system

Ideally, a zone with high susceptibility and risk of landslide should be prevented from any development, such as for the development of housing and infrastructures. Unfortunately, it is quite often that this particular zone has been developed as dense-settlement of housing and infrastructure areas, and the relocation of people living in this particular area to the saver zone cannot be conducted due to some socio-economical constraints. Thus, the development of a landslide early warning system becomes very critical to protect the people living in that particular landslide risk area.

Therefore, it is important that the landslide early warning system should be performed with a low cost and simple technology that can be easily understood, operated and maintaned by the local community, such as suggested by Karnawati et al in [9] and by Fathani et al. in [10]. Indeed, such system should be the integration of technical and social systems. The technical system consists of simple extensometers, i.e. to monitor the ground-surface movement with the accuracy of 0.1 mm, and a rain-gauge used to monitor the critical rainfall which can induce landsliding with the accuracy of 1 mm, fig. 8. Meanwhile the social system is supported by the network of selected individuals who are assigned by the local community in the village as the member of community task force for disaster mitigation and management (as discussed in Section 5.2), and this team is responsible for the installation, operation and maintenance of the technical system. For instance, the system which has been installed in two selected sites in Central Java consisted of 5 extensometers and 1 rain-gauge for each site with an area of about 50 to 75 hectares, and supported by 15-man-power selected from 5 housing groups in each site. To provide an early warning, an alarm is connected to each piece of equipment in the system and the alarm is automatically set to be "on" when critical rainfall which can induce landsliding and/or critical conditions of slope movement occurs. To set the alarm to be on at the appropriate time, all of those pieces of equipment are generated by dry batteries and/or solar energy which can work effectively during heavy rainfall when the electricity power does not work properly. The village action plan (including the contingency plan) for disaster prevention and response program is accordingly developed by this task force team. Obviously, one of the most important programs to guarantee the effectiveness in implementing this early warning system in the village is public education and evacuation drill such as that illustrated in fig. 9. Those programs need to be conducted regularly to improve the awareness and readiness of local community for any possible landslide disaster. In fact, this simple early warning system has successfully saved 35 families from a landslide which occurred in Kalitelaga village at Banjarnegera Regency on November 7, 2007, such as reported by Karnawati *et al.* in [9].

Figure 8: Extensometer (left) with the dry batteray (center) and alarm siren (right) in [9].

Figure 9: Evacuation drill (left) and the coordination with the local rescue team (center and right).

6 Conclusion

Despite the natural conditions which are susceptible for landslide occurrence, potential landslide disaster in Central Java can be reduced with more emphasis on the management of social vulnerability. Indeed, this disaster risk reduction effort needs to be supported by both technical and social approaches. The technical approach is required to provide facilities for immediate access in landslide mitigation/prevention, whilst the social approach should mainly directed for the capacity development of the local community in the village. Accordingly, an integrated approach of both a technical and social system should be conducted to establish an effective community early warning system. Through all of these strategic approaches, community resilience in disaster-prone areas at village level can be effectively improved. It is also very crucial to especially address the traditional wisdom and indigenous-simple technology in both technical and social approaches for disaster risk reduction. Finally, it should be admitted that the success of disaster risk reduction at the local or at village level is the key-success for the disaster risk reduction program at a national level.

Acknowledgements

The authors would like to express their gratitude to the DelPHe Project funded by the British Council for providing financial support to carry out the action research presented in this paper. Special acknowledgement is also directed to the WCRU Program of Gadjah Mada University for providing the sponsorship to participate in this Disaster Management 2009 Conference.

References

[1] G.S.G.S., *Geology of Central Java*, Sheets S B 48, 49, 50, SC 49, 50. 2nd ed. Photographed by War Office: USA, 1944.

[2] Surono, Toha, B. & Sudarno I., *Geology of Surakarta-Giritontro*, Geological Research and Development Centre: Indonesia, 1992.

[3] Sampurno & Samodra, H., *Geology of Ponorogo*, Geological Research and Development Centre: Indonesia, 1997.

[4] Karnawati, D., Ibriam, I., Anderson, M.G., Holcombe, E. A., Mummery, G.T., Renaud, J.P & Wang, Y., An initial approach to identifying slope stability controls in Southern Java and to providing community-based landslide warning information. *Landslide Hazard and Risk*, eds. T. Glade, M.G. Anderson & M. J. Crozier, John Wiley and Sons: New York, pp. 733-763. 2005.

[5] Karnawati, D. & Fathani, T. F., Mechanism of earthquake induced landslides in Yogyakarta Province, Indonesia. *The Yogyakarta Earthquake of May 27, 2006*, eds. D. Karnawati, S. Pramumijoyo, R. Anderson & S. Hussein. Star Publishing Company Inc.: Belmont, CA., pp. 8-1 to 8-8, 2008.

[6] Andayani, B., Karnawati, D. & Pramumijoyo, S., Institutional frame work for community empowerment towards landslide mitigation and risk reduction in Indonesia. *Proc. of the 1st World Landslide Forum*, Global Promotion Committee of the Int. Program on Landslide (IPL) – ISDR: Tokyo, pp. 57-59. 2008.

[7] Karnawati, D. & Burton, P.W., *Seismicity and Landslide Research towards Public Empowerment for Hazard Preparedness; First Year Annual Report of Development Partnership in Higher Education*

(DelPHE) – the British Council, Geological Engineering Department, Faculty of Engineering, Gadjah Mada University-School of Environmental Sciences, University of East Anglia: Indonesia and UK. 2008.
[8] Karnawati, D., Pramumijoyo, S., Andayani, B. & Burton, P.W., Earthquake & landslide hazard mapping for community empowerment. *Proc. of the 51st Annual Meeting Assoc. of Engineering and Environmental Geologist*: New Orleans, 2008.
[9] Karnawati, D., Fathani, T.F., Sudarno, I. & Andayani, B., Development of community-based landslide early warning system in Indonesia. *Proceeding of the 1st World Landslide Forum*, Global Promotion Committee of The Int. Program on Landslide (IPL) – ISDR: Tokyo, pp. 305–308. 2008.
[10] Fathani, T.F., Karnawati, D., Sassa, K. & Fukuoka, H. Development of landslide monitoring and early warning system in Indonesia. *Proc. of the 1st World Landslide Forum*, Global Promotion Committee of The Int. Program on Landslide (IPL) – ISDR: Tokyo, pp. 195–198.

Erosion of forestry land: cause and rehabilitation

T. Ogawa[1], Y. Yamada[2], H. Gotoh[2] & M. Takezawa[2]
[1]*Forest Survey Office, Japan*
[2]*Department of Civil Engineering, College of Science & Technology, Nihon University, Japan*

Abstract

Forests cover 70% of the total land area of Japan. Forest lands within Japan are prone to landslides because weakly resistant geological units are eroded by water flowing down steep slopes that are subjected to annual rainfall amounts that are 2.5-times the global average. The environmental effects of deforestation impact upon atmospheric pollution, wildlife, the hydrological cycle, water resources, soil erosion, and the occurrence of landslides. To mitigate disasters that occur upon forestry land, it is important to forecast landslide development and plan for the provision of remedial measures during disaster rehabilitation. This paper describes the causes of the erosion of forestry land and methods of disaster rehabilitation via a case study of the upper reaches of the Tama River, Japan, which is a national park and an important water resource for the Tokyo Metropolitan area. The causes of erosion of forestry land within the upper reaches of the Tama River are classified as one of the following: shallow landslides related to the loss of under-story vegetation, collapse of steep slopes, damage related to the consumption of vegetation by wildlife, and debris flows that occur during periods of torrential rain. In recent times, heavy rains over eroded forestry land within the upper reaches of the Tama River have produced muddy river water due to the erosion and degradation of mountain slopes. In addition, grazing by Japanese deer has destroyed many trees within the upper reaches of the Tama River, and the torrent bed within this area, previously planted with Japanese horseradish, was lost during a debris flow. In this paper, we describe anti-erosion measures undertaken for disaster rehabilitation of wasted forestry land, including timber-thinning methods and the control of wildlife numbers.

Keywords: erosion of forestry land, deforestation, disaster rehabilitation, landslide.

1 Introduction

It is fair to say that a human life is created by the green of the forest that brings about a mild climate, saves a water resource, and serves the coexistence of animals and plants. The forest is a stable system that can sustain nature. About 70% of the total land area of Japan is forested. Japan is blessed with the most abundant forest resources of any country in the world. The forests of Japan were depleted during the Second World War, but Japanese red cedar (*Cryptomeria japonica*) and Hinoki cypress (*Chamaecyparis obtuse*) were replanted during the post-war period at the demand of the Japanese Government. Plantation forests currently comprise about 40% of the total forests in Japan, but such wood is in poor demand because foreign lumber can get purchased cheaply. The cost of a log of Japanese red cedar or Hinoki cypress is just 33% or less of the price 25 years ago. The rate of self-sufficiency of forests in Japan is less than 20%; consequently, the average age of forestry workers increases and the next generation of forestry owners have moved to the city because forestry is not a desirable career. Hence, the management of forests in Japan has been neglected in recent times, leading to erosion and poor water conservation. In this paper, we discuss the causes of eroded forest land and possible rehabilitation measures.

2 Forest land in Japan

The forested area of Japan has decreased with increasing population and the development of agricultural fields. The cultivated acreage of forest was about 8.62 million ha for about 7,000,000 people in 930 AD, increasing to 54 million ha for 121,000,000 people in 1990: an increase in persons per 1 ha of cultivated acreage from 8.1 to 22.4. Temporal trends in population and cultivated acreage are shown in Table 1 (Iketani [1]).

Table 1: Temporal trends in cultivated acreage and population within Japan.

Year	Cultivated acreage (million ha)	Population (million)	Persons per hectare
930	8.62	7	8.1
1450	9.46	10	10.5
1600	16.35	19.6	12.0
1720	29.70	31	10.4
1874	30.50	34	11.1
1990	54.00	121	22.4

The sizes of areas of protected forest in Japan in 2004 are shown in Table 2. The 'Other' category in the table includes shifting-sand prevention forest, windbreak forest, flood-damage prevention forest, tidal wave and salty wind prevention forest, drought-prevention forest, snow-drift prevention forest, fog-inflow prevention forest, snow-avalanche prevention forest, rock-fall prevention forest, fire protection forest, fish-breeding forest, navigation landmark forest, public health forest, and scenic-site conservation forest. Headwater conservation forest comprises 68.4% of protected forest, while soil run-off prevention forest comprises 21.5%, as shown in Table 2 (JFS [2]).

Table 2: Land areas of different types of protected forest.

Classification	National forest	Non-national forest	Total	Ration
Headwater conservation forest	4,228	3,216	7,444	68.4%
Soil run off prevention forest	935	1,404	2,399	21.5%
Landslide prevention forest	19	37	56	0.5%
Other	458	590	1,048	9.6%

(unit: thousand ha)

Felled forestry land was 19,830 ha; the total area of landslides per 100 ha was 2.38 ha; and the total area of landslides in areas of mature forest (190,328 ha) was 1.17 ha per 100 ha. Differences in landslide development between areas of mature forest and felled areas are shown in Table 3 (JSECE [3]). The total landslide area per 100 ha of felled land is about 2-times that of mature forest areas, as shown in Table 3. Therefore, one of the causes of landslide development is deforestation.

Table 3: Relations between areas of felled forest and landslide development.

Kind	Area (ha)	Landslide (places)	Landslide (ha)	Landslide area (100 ha)
Tree-grown area	190,328	11,286	2,277	1.17
Cut-over land	19,830	2,377	398	2.38
Total	210,158	13,663	2,625	1.25

Trees and plants in general affect the hydrological cycle in a number of significant ways. Therefore, the presence or absence of trees can change the quantity of water upon the land surface, within the soil or groundwater reservoir, and in the atmosphere. Deforestation generally increases the rate of soil erosion by increasing the amount of runoff and reducing the protection of the soil afforded by tree litter. Forestry operations themselves also increase erosion via the development of roads and the use of mechanized equipment. A further cause of landslide development is the grazing of wild animals. In some cases, treeless hills result from the grazing upon growing herbage of wild animals such as deer, monkey, and bear. The surface of the earth is then exposed and washed away during rainfall. Tree roots act to bind soil between the roots and between the roots and underlying bedrock if the soil is sufficiently shallow. The risk of landslides is therefore increased when trees are removed from steep slopes with shallow soil and when subjected to the grazing of wild animals. The following case study provides an example of the way in which the rehabilitation of eroded forest land is planned and executed in Japan.

3 Case study

The rehabilitation of eroded forest land is planned for the area around the Sakasa River, a branch of the Tama River, as shown in Figure 1.(JT[4]) The study area of the current investigation comprises 131 ha of fast-moving river and hillsides within the 220-ha basin catchment of the Sakasa River. The Sakasa Basin contains 198.07 ha of forest that is designated a protected area with the status of 'protection for headwater conservation'.

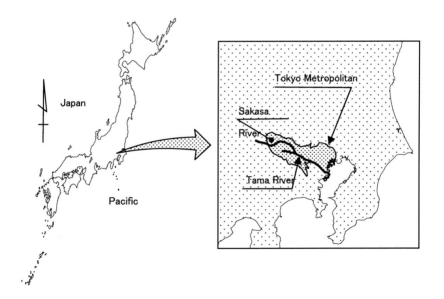

Figure 1:   Location map of the study area.

The study area is the mountainous region between Kawanori Mountain, with an altitude of 1353 m, and Honnita Mountain, with an altitude of 2245 m. The upper reaches of the Sakasa River flow to the south, while the middle reaches flow to the west following a 90° change in flow direction at an altitude of 790 m related to a geological rift. The topography of the upper Sakasa River formed from crustal movements, and large-scale landslides occur at 3 sites in the upper river. Hillsides in this area are steep, with an average slope of 36°. Eroded subvertical cliffs occur at the levee foot of mountain streams in particular, and the slopes of the lower parts of such cliffs are in excess of 60°. The bed slope of the Kawanori River where it joins the Sakasa River is 10°; this section of the river forms a V-shaped valley with numerous falls of 50–60 m. The average slope of the riverbed of the upper Sakasa River is 16°; this section of river contains many falls of 2–10 m, and rocks are exposed along the majority of the rivercourse. There are three main river branches within the Sakasa Basin. Most areas within the river basins are 20–45 ha in size and many bed slopes are in excess of 20°. The topography of the Sakasa River consists of six distinct regions, as shown in Table 4 and Figure 2. Figures 3 and 4 are the erosion of forest land and the devastated land by grazing of Japanese deer in the Kawanori Mountain of Tokyo Metropolitan.

The elongation ratio (E) is given by the equation $E = (2/L)\sqrt{(A/\pi)}$, where L is the river length (m), A is area (ha), and π is the circular constant. The geology of the study area consists of Late Jurassic sandstone underlain by Middle Jurassic shale. The sandstone and shale is overlain by a layer of loam in mountainous areas; the loam is weak and commonly deformed.

Temperature and precipitation data for altitudes of 550 m, 1000 m, and 1363 m are shown in Table 5. The maximum recorded rates of rainfall are 71 mm/hour (1991/8/20), 347 mm/day (2001/9/10), and 634 mm/3 days (2001/9/10).

Cryptomeria japonica and *Chamaecyparis obtuse* planted over the past 30–50 years now covers about 60% of the total forest area across the study site. A classification of the forest area on the basis of tree type is shown in Table 6, while eroded forest land is classified in terms of three types of eroded hillsides and torrent-erosion, as shown in Table 7. Eroded forest land is classified in terms of rainfall, intensity of felling, and damage related to grazing by

Table 4: Characteristics of the topography of the Sakasa River.

Part	Area (ha)	Length (m)	Height ratio	Bed slope (%)	Slope angle (deg)	Elongation ratio
Lower	35.85	720	160	22	41	
Midstream	10.63	540	90	17	38	
Upper	85.64	1390	460	33	35	0.75
1 stream	19.84	570	290	51	38	0.88
2 stream	22.72	630	330	52	37	0.85
3 stream	45.32	900	360	40	35	0.84
Total	220.00	2650	710	28	37	0.63

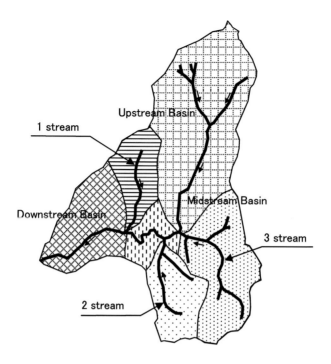

Figure 2: Six-part topographic division of the Sakasa River catchment.

Figure 3: Erosion of forest land.   Figure 4: Devastated land.

Japanese deer. In recent years, Japan has been struck by severe typhoons and torrential rain, such that the amount of eroded forest land has increased due to a combination of these rainfall conditions, weak soil, and the grazing of Japanese deer. The main cause of erosion of forest land in the study area is grazing by increasing numbers of Japanese deer that eat the buds and roots of trees. Rehabilitation of the eroded forest land requires the control and management of an appropriate population of Japanese deer. Measures to prevent run-off include hillside works such as the conservation of vegetation within devastated lands and torrent works such as the protection of valleys that contain unstable soil. Figure 5 and 6 are the forest of *Cryptomeria japonica*.

Table 5: Temperature and precipitation data for different altitudes.

	Altitude 530 m	Altitude 1000 m	Altitude 1363 m
Average temperature	11.8°C	9.0°C	6.8°C
Volume of temperature Warming	90.1	67.7	51.6
Volume of temperature Coldness	-8.5	-19.7	-30.0
Classification of forest zone	Temperate zone, evergreen and broadleaf forest	Temperate zone and fallen leaf forest	Cold zone and fallen leaf forest
Annual precipitation	1595 mm	1971 mm	2259 mm

Figure 5: Forest of *Cryptomeria japonica*.

Figure 6: *Cryptomeria japonica*.

Table 6: Classification of forest areas on the basis of tree type.

Kind of tree	Area
Cryptomeria japonica and Chamaecyparis obtuse	60.63 ha (27%)
Cryptomeria japonica	28.45 ha (13%)
Chamaecyparis obtuse and Broad leaved trees	33.70 ha (15%)
Chamaecyparis obtuse	3.30 ha (2%)
Quercus mongolica var. grosseserrata and Broad leaved trees	49.80 ha (23%)
Quercus mongolica var. grossesrrata and Larix leptolepis	3.75 ha (2%)
Quercus mongolica var. grossesrrata	5.57 ha (3%)
Quercus serrate and Broad leaves trees	16.05 ha (7%)
Others broad leaves trees	12.05 ha (5%)
Tsonga sieboldii	4.10 ha (2%)
Pinus densiflora	0.80 ha (0%)
Larix leptolepis	1.80 ha (1%)
Total	220.00 ha (100%)

Table 7: Classification scheme of eroded forest land.

Classification	Form
Eroded hillside (1)	Outflow of surface soil by loss of under-story vegetation
Eroded hillside (2)	Slope failure of steep slope
Eroded hillside (3)	Harm of deer's food
Eroded torrent	Outflow zone of soil by torrent bed gradient

Table 8 describes the characteristics of different types of hillside and torrent works in terms of: (1) type of works, (2) effectiveness, (3) durability, (4) value for money, (5) ease of undertaking the work, (6) ease of transportation of materials, (7) protection of the landscape, (8) combats grazing by Japanese deer, and (9) overall evaluation. The meanings of the symbols used in the figure are as follows: ◎: very good performance; ○: good performance; △: poor performance; and ×: very poor performance. Hillside work is classified as either technical hillside work or hillside-seeding work. Table 8(a) assesses soil-retaining works as part of technical hillside works. Basket-retaining works (A) perform well in both sides of the water-channel work, while log-piling work (C) performs well in soil-retaining works Table 8(b) assesses water-channel work as part of technical hillside work; the wire net (F) shows the best performance.

Table 8(a): Technical hillside works (soil-retaining works).

(1)	(2)	(3)	(4)	(5)	(6)	(7)	(8)	(9)
A	○	○	○	○	◎	◎	-	◎
B	○	○	○	△	△	△	-	△
C	◎	△	△	○	○	◎	-	◎
D	◎	△	○	○	○	◎	-	○

A: Basket retain work, B: Wire basket work, C: Log piling work,
D: Wood steel wall.

Table 8(b): Technical hillside works (water-channel works).

(1)	(2)	(3)	(4)	(5)	(6)	(7)	(8)	(9)
E	○	○	○	△	◎	○	○	○
F	◎	◎	○	○	○	○	◎	◎
G	○	○	△	△	△	○	×	△
H	○	○	△	△	△	△	○	△

E: Sand bag, F: Wire net, G: Sodded channel, H: Corrugated metal pipe.

Table 8(c) provides an assessment of different wattle fence and linear sodding works as part of hillside seeding works. Log wattle fences (J) and log linear sodding (M) provide good performance because thinning lumber can be used.

Table 8(c):   Hillside seeding works (wattle fence and linear sodding works).

(1)	(2)	(3)	(4)	(5)	(6)	(7)	(8)	(9)
I	○	△	○	△	△	◎	○	△
J	○	◎	○	○	○	◎	○	◎
K	○	○	○	○	○	○	○	○
L	○	○	○	○	○	△	○	○
M	○	△	○	○	○	◎	○	◎
N	○	△	○	○	○	○	○	○

I: Wicker work, J: Log wattle fence, K: Steel linear sodding work,
L: Wattle fence with vegetation, M: Log linear sodding work,
N: Linear sodding work with vegetation.

Table 8(d) provides an assessment of covering works as part of hillside seeding works. In this case, natural fiber mat (R) provides the best performance.

Table 8(e) provides an assessment of covering works to protect vegetation from grazing by Japanese deer. Combined works involving a thick layer wire net and SHIKATTO (wire netting bounded by coil springs) works [5] provides the best results, where the SHIKATTO work provides protection from grazing by deer. Figure 7 is the SHIKATTO work and Figure 8 is the Japanese deer [6]. Table 8(f) provides an assessment of torrent work for protecting valleys that contain unstable soil. Torrent beds previously planted with Japanese horseradish have been repeatedly lost to debris flows.

Figure 7:   SHIKATTO work [5].   Figure 8:   Japanese deer (SHIKA) [6].

Table 8(d): Hillside seeding works (covering works).

(1)	(2)	(3)	(4)	(5)	(6)	(8)	(9)
O	◎	○	○	△	△	○	△
P	○	△	◎	◎	◎	△	△
Q	◎	○	△	△	△	△	△
R	◎	○	◎	○	○	△	○
S	○	○	◎	○	○	△	○

O: Fagot, P: Straw mat with seed, Q: Vegetation mat with thick top soil,
R: Natural fiber mat, S: Chemical fiber mat.

Table 8 (e): Hillside seeding works (covering works: ward off Japanese deer).

(1)	(2)	(3)	(4)	(5)	(6)	(8)	(9)
T	○	○	○	○	○	◎	◎
U	○	○	○	○	○	◎	○
V	◎	○	○	○	○	◎	◎
W	◎	○	○	○	○	◎	◎

T: Wire net together, U: Diamond wire net, V: Thick layer wire net,
W: SHIKATTO work.

Table 8 (f): Torrent works.

(1)	(2)	(3)	(4)	(5)	(6)	(7)	(9)
I	◎	○	○	○	○	◎	◎
II	◎	△	△	○	○	◎	○
III	○	○	△	△	△	○	×
IV	○	○	○	○	△	○	○
V	○	◎	△	△	×	△	×

I: Steel crib structure, II: Wood crib structure, III: Cellular structure,
IV: High energy absorber fence, V: Concrete structure.

Torrent works using trench soils and secondary products that are easily transported are selected for crib dams; the wood crib structure is the structure that is mainly used. The steel crib structure is best in downstream areas and at river junctions. Figure 9 is a torrent zone at the upstream of the Sakasa River and Figure 10 is a crib dam using thinning timbers of a forest.

Figure 9:   A torrent zone.   Figure 10:   Crib dam.

4 Conclusions

The causes of the erosion of forest lands include intense rainfall events in recent years, deforestation, and increases in wildlife numbers. The effects of global warming are increasing year-by-year: evaporation increases with a warming climate; the average global precipitation increases; soil moisture is likely to decline in many regions; and intense rainstorms are likely to increase in frequency. Deforestation within a basin causes water- and wind-derived erosion, a decline in soil fertility, and the development of landslides. It is known that the roots of Japanese red cedar and Hinoki cypress decay within 7 years of felling. It is therefore unreasonable to expect prolonged resistance to landslide development in areas of felled trees because the roots decay over time. The process of landslide development following the grazing of wild animals involves the removal of surface soil, gullying, and rill formation. The number of Japanese deer in the study area has been increasing rapidly because the area provides a favorable environment for the deer. The population of Japanese deer in this area has increased by 660% over the past 10 years. The rehabilitation of eroded forest land must involve both forestation measures and measures to control and manage the wildlife population at a suitable level.

The Forest Improvement Plan for the devastated area is to be carried out comprehensively via the Forestation Project and the Protected Forest Improvement Project. The Forest Improvement Plan is concerned with the growing conditions of trees, the existence of under-story vegetation, crown density, the ratio of forms, the ratio of yield, etc. The plan involves the growth of various types of under-story vegetation, the development of an ecologically multi-storied forest with sufficient ground litter, sufficient intensity of illumination under the tree crown, and enhancing the growth rate of trees by periodically carrying out forest improvement works. Such works enhance public services such as water conservation and the prevention of run-off by afforestation, and the best environment is created for wildlife.

Acknowledgement

A part of this research was conducted with the financial support of the Tokyo Metropolitan Government. Authors thank cooperators of this survey.

References

[1] H. Iketani (1999) Disaster of Debris Flow, Iwanami Series, pp.29.
[2] Japan Forestry Society (2005) Forest Handbook, pp.9.
[3] Japan Society of Erosion Control Engineering (1999) Movable Events of Soil at Slope, Sabo Course Vol. 3, pp.169.
[4] Japan Technology Co. Ltd. (2005) General Survey Report of Forest Conservation Project in the Sakasa River, pp.1–197.
[5] Asuka Landscape Architecture Co, Ltd. (2010) SHIKATTO tree planting work, http://www.asuka-la.co.jp/nmn/index.html
[6] Wikipedia (2010) Sika deer, http://ja.wikipedia.org

Landslide in a catchment area of a torrent and the consequences for the technical mitigation concept

F. J. Riedl
Austrian Federal Service for Torrent, Erosion and Avalanche Control, District Office "Middle Inn Valley", Austria

Abstract

The focus of the following lies on the identification of the practical solutions by taking into account the theoretical background. A translational landslide occurred on October 2008 within an area of about 2.5 ha on the topographical left side of the torrent "Wattenbach" in Tyrol/Austria.

In the summer of 1965 a large debris flow event of the torrent "Wattenbach" happened and the underlying city of Wattens was completely destroyed. After this event several technical protection measurements were implemented to guarantee a certain factor of safety for the city of Wattens.

After the landslide event of 2008, the most important question concerned the interaction of the torrent "Wattenbach", the landslide and which reaction could be expected by a flooding event in the future.

To obtain certain quantitative and qualitative data, several analyses (modelling of the landslide by regarding different scenarios, laser scans, field works) were implemented and some of them are still going on.

Keywords: landslide, debris flow, torrent, Austrian Federal Service for Torrent, Erosion and Avalanche Control, technical protection measurements, natural hazards.

1 Introduction

1.1 Austrian Federal Service for Torrent, Erosion and Avalanche Control, District Office "Middle Inn Valley"

The main tasks of the Austrian Federal Service for Torrent, Erosion and Avalanche Control are divided into hazard zone planning (risk prevention), planning of technical mitigation measures against natural hazards, building constructions works and expertises for the public authorities.

The District Office "Middle Inn Valley", situated in Innsbruck, is responsible for the torrent, avalanche, rockfall and erosion protection in the two districts Innsbruck-Town and Innsbruck-Country. The total area covers 210.000 ha whereby only 15% of them can be used for permanent settlement.

There are also 256 torrent catchment areas and 264 harmful avalanche tracks which endanger the permanent settlement. In the year 2009 the monetary investments were about 5.0 million € in the preventive technical, forestry and soil-bioengineering measurements.

2 Landslide event on October 2008, city of Wattens, district of Innsbruck-Country

2.1 Kinematic analysis of the landslide in the catchment area of the torrent "Wattenbach"

The landslide in the catchment area of the torrent "Wattenbach" occurred on October 2008 on the topographical left side of the torrent. For analysing purposes, a field work was conducted to obtain data about the geology, the geomorphology, the level of the mountain water, the surface runoff and the initial structural situation (brittle and ductile deformation, foliation, etc).

2.1.1 Geology
The geology is defined by a fine foliated phyllite, the so-called "Innsbrucker Quarzphyllite". In general, the northeast-exposed hillsides are ancient deep seated gravitational landslides which are nowadays in a nearly firm stage [1]. From the structural and tectonical point of view the phyllite was deformed by several different phases from D1-D4 [2]. Actually, a steady influence of the

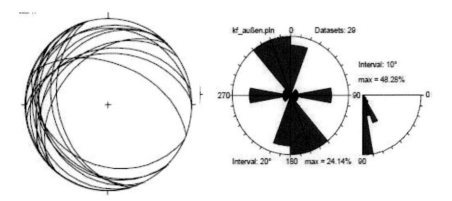

Figure 1:   Orientation of the foliation and the brittle cracks within the landslide [3].

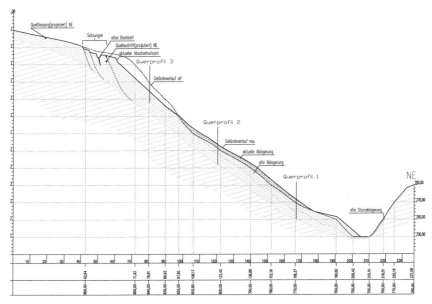

Figure 2: Geomorphological mapping and a profile through the active translational landslide [3].

active brittle sinistral Inntal fault can be examined which causes a negative structural influence on the stability of the preliminary deformed phyllite.

In this specific case the dip of the foliation diversifies, which is caused by a singular rotation, from south to west. Furthermore a nearly constant dip of the foliation to the southwest can be recognized. The brittle deformations within the landslide are characterized by steep northwest-southeast orientated cracks [3].

2.1.2 Geomorphology
From the kinematical point of view the landslide is a translational slide [4], partly within the solid rock, with a depth of the slip plane of about 10-20 m. As mentioned before the active landslide is part of an enlarged ancient deep seated gravitational landslide system. The torrent "Wattenbach" has eroded the convex front of the moving part during the last centuries and since the catastrophically debris flow event of 1965, the toe of the slope has become precipitous. The main scarp dips 50-70° to the northeast and on the topographical right side of the boundary zones the dip of the scarp rotated to the north.

2.1.3 Hydrogeological and hydrological runoff
There are two diffuse hydrogeological mountain water zones, one between 850-860 m a.s.l. and the other one between 810-820 m a.s.l. The hydrological surface

Figure 3: Debris flow event in August 1965 and the damages in the middle part of the torrent and the height of the water level by the bridge in the city of Wattens (above).

Table 1: Results of the slope stability analysis.

Szenarien	Kombination 1	Kombination 2	Kombination 3
a1	1,0	1,0	1,0
b1	1,0	1,0	0,90
b2	0,83	0,81	0,83

Tabelle 1: Ergebnisse für die Berechnungen mit Geländeprofil 1

Szenarien	Kombination 1	Kombination 2	Kombination 3
a2	1,21	1,14	1,10
c1	1,10	1,10	1,08
c2	0,98	0,98	1,05
d1	1,21	1,21	1,10
d2	1,21	1,21	1,10

Tabelle 2: Ergebnisse für die Berechnungen mit Geländeprofil 2

szenarios

a1 development of the landslide topographic profile 1
a2 development of the landslide topographic profile 2
b1 influence of the mountain water zone 1 by profile 1
b2 influence of the mountain water zone 2 by profile 1
c1 decrease of the riverbed 5,0m profile 2
c2 decrease of the riverbed 10,0m profile 2
d1 increase of the riverbed 5,0m profile 2
d2 increase of the riverbed 10,0m profile 2

runoff is marked by several small gullies with some initial erosion. Local technical mitigation measures have been implemented by the Austrian Federal Service for Torrent, Erosion and Avalanche Control Service, District Office Middle Inn Valley on a tributary to the "Wattenbach".

The catchment area of the main torrent "Wattenbach" is 74 km² with a peak runoff of about 90m³/s with a regarded repeat period of time of 150 years. On the base of the hazard zone planning, the expected bed load is about 160.000m³. The debris flow event of 1965, where large areas of the city Wattens were destroyed, the measured bed load was about 85.000 m³ [5].

The main drinking water spring of the city Wattens is above the active landslide on 920 m a.s.l. and the remaining water was currently flowing uncontrolled into the active moving zones.

2.2 Geotechnical investigations and modelling

The main tasks of the geotechnical investigations were to examine the development of the landslide, the role and importance of the mountain water level and the development of the landslide by decreasing and increasing the riverbed of the torrent "Wattenbach". The numeric modelling of the landslide, by regarding the topics, mentioned above, was done by an extern consulting engineering company with the finite element software PHASE2, Rocscience Inc. The results of this modelling should be the base for the further technical mitigation concept.

By analyzing the different scenarios (two topographical profiles and the influence of the two zones of the mountain water levels) the results of the slope stability analysis are as follows.

The main conclusion of these geotechnical investigations and the results of the varying scenarios were the fact that a decrease of the riverbed of the torrent "Wattenbach" with more than 10,0 m would lead to a slope failure (c2). This

Abbildung 42: Grenzzustand des Hanges (Deformationen farblich dargestellt)

Figure 4:   Modelling of the landslide and the deformation by regarding the different scenarios.

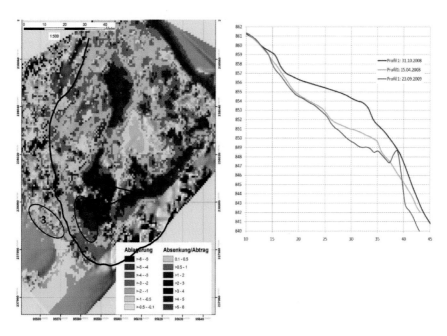

Figure 5:   Difference in height inside the landslide and the distinctive depression in the upper part [7].

induces to a technical mitigation concept whereby the riverbed has to be fixed against depth erosion. An increase of the riverbed results in a minor increase of the factor of safety (d1-d2). Nevertheless the mountain water level is fundamental and has to be considered for the further mitigation planning [6].

2.3 Laser scanning of the landslide

In the alpine regions the original use of the laser scanning is based on the examination of the snow cover. The scanning of a surface of a landslide and the obtained experiences for the future was one of the defined tasks. A cooperation partner, the Federal Research and Training Centre for Forests, Natural Hazards and Landscape, has fulfilled on base of a cooperation contract the laser scanning of this landslide. The scanning was done with the Scanner LPM98-2K by the company Riegl with the highest cancelation and a projected spot spacing of 0,75m [7].

The focus of the laser scanning investigation was to obtain compressions and depressions within the active zones and further detailed information about the kinematical movements. It is also important to point out the fact that the technical mitigation achievements should be quantified after finishing the construction works for a defined period of 3-4 years. The reference measurement has already been created after the event 2008 and the main confiding at this time were the enormous depressions on the upper part of the landslide up to 6,0m.

3 Technical mitigation concept

3.1 General conspectus

On the base of the several investigation results, the technical mitigation concept was according to them.

The main conclusion of the slope simulation was the fact that the riverbed should be fixed and consolidated to obtain slope stability. Due to the induced depth erosion, caused by the debris flow event in den "Wattenbach" in the summer of 1965, a collapse of the slope stability could occur and additionally bed load material could be transported up to the city of Wattens.

3.2 Consolidation of the riverbed of the torrent "Wattenbach"

For the stabilisation and consolidation process of the riverbed of the torrent "Wattenbach", at least 13 check dams with a certain debris flow section will be constructed to avoid depth erosion during a debris flow event. The dimension of the check dams will be designed on the peak runoff of 90m^3/s. The needed height of the check dams is about 3-4m, the concrete cubature is about 300m^3 and the steel demand is about 8,5 tonnes per check dam. The toe of the landslide should be firm up by stabilizing the riverbed through these check dams.

122 Safety & Security Engineering

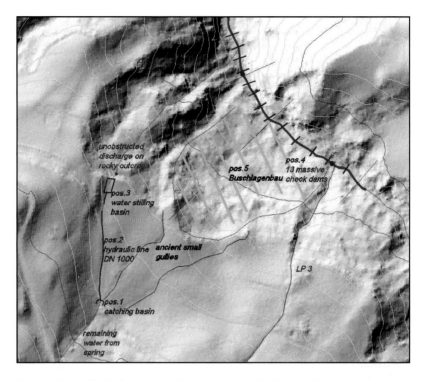

Figure 6:   Technical mitigation concept at the basis of the investigations.

3.3 Hydrological surface runoff of the small gullies

The small gullies beyond the landslide, which infiltrated into the instable slope, will be displaced with an earth-covered hydraulic line DN1000 to the rocky outcrops on the topographical left side of the landslide.

To decrease the kinematic force of the water, a stilling basin will be constructed on base of the hydraulic dimension. Due to the rocky outcrops the surface runoff of the small gullies can then be discharged unobstructed on them.

3.4 Soil-bioengineering measurements

The diffuse characteristic of the two mountain water zones within the landslide will be conceived by soil-bioengineering measurements, the so-called "Buschlagenbau" [8]. Differential salix are used for the "Buschlagenbau" and they will be orientated to the topographical left rocky outcrops, to achieve a two-dimensional drainage effect. Another aspect of the "Buschlagenbau" is a small area stabilisation of the upper soil-complex. These soil-bioengineering measurements are the highly recommendable measurements due to uncontrolled diffuse water flow. The main mechanical movements of an unstable slope are in general caused by such diffuse water flows.

3.5 Perspectives

To obtain the expected achievements of this technical mitigation concept it is mandatory to make further investigations in analyzing the landslide with the laser scanning after finishing the construction works. The quantitative and qualitative movements of the investigation points will give an explanation of the landslide and a confirmation if the technical measurements worked as successfully as expected.

References

[1] Hermann, S., Tiefreichende Großhangbewegungen im Kristallin der Niederen Tauern, Ostalpen. –Verbreitung, Typen und ihr Einfluss auf die Morphogenese alpiner Täler. *Geoforum Umhausen (GFU)*, **1**; Innsbruck, 1999

[2] Rockenschaub, M., Kolenprat, B., Frank, W., The tectonometamorphic evolution of Austroalpine units in the Brenner area (Tirol, Austria) – new geochronological implications. *Tübinger Geowissenschaftlichen Arbeiten, Series A*, Vol. **52**, pp. 118–119, 1999.

[3] i.n.n., Rutschung Eggerbachl - Bewertung des Ist-Zustandes und Abschätzung der Auswirkungen auf den Hochwasserabfluss im Wattenbach. *Intern Report*, Innsbruck, 2009.

[4] Varnes, D. J., *Slope Movement Types and Processes – in: Special Report 176: Landslides: Analysis and Control*, (R. L. Schuster and R. J. Krizek, eds.), TRB, National Research Council, Washington D.C, 1978.

[5] Forsttechnischer Dienst für Wildbach- und Lawinenverbauung, Wattenbach. *Kollaudierungsoperat 1993 für die Baujahre 1965-1991*, Technischer Bericht, pp. 21, Innsbruck, 1993

[6] GEC ZT GmbH, Rutschung Wattenbach/Eggerbach. *Intern Report*, Innsbruck, 2009.

[7] Federal Research and Training Centre for Forests, Natural Hazards and Landscape Department Natural Hazards and Alpine Timberline Unit Water Balance in Alpine Catchments. Kurzinformation zur Massenbewegung Wattental, Laserscanning, *Intern Report*, Innsbruck, 2009.

[8] Schiechtl, M., Stern, R., *Handbuch für naturnahen Erdbau*, Österreichischer Agrarverlag: Wien, pp. 92-96, 1992.

Slope instability along some sectors of the road to La Bufadora

J. Soares[1], C. García[2], L. Mendoza[3], E. Inzunza[1],
F. Jáuregui[4] & J. Obregón[5]
[1]*Engineering Faculty of the Universidad Autónoma de Baja California UABC in Ensenada, Baja California, Mexico*
[2]*Instituto Municipal de Investigación y Planeación, Mexico*
[3]*Centro de Inv. Científica y de Ed. Superior de Ensenada CICESE, Mexico*
[4]*Social and Administrative Sciences Faculty of UABC, Mexico*
[5]*Subheadship of the Civil Protection County Agency, Ensenada, Baja California. Mexico*

Abstract

La Bufadora is a famous water sprout and a tourist spot located about one hour drive south of the city of Ensenada, in the state of Baja California, Mexico. The route to La Bufadora is a narrow curvy road with unstable slopes. On a 4m. stretch, the road continuously subsides, probably due to a sector of the Agua Blanca Fault that crosses the Punta Banda Peninsula. Swarms of small earthquakes have been located in this same section by RESNOM since 1997 (CICESE, RESNOM Online catalog. http://sismologia.cicese.mx/resnom/catalogo/datain.php) and unstable rocks presuppose a risk that is manifested in the occurrence of slips. In this paper we will show a slope classification made according to the need of attention as well as their remedial options. We document new evidence of an anthropogenic impact: a high pressure of tourist housing development in several places of the peninsula. We registered and identified the causes for the slope instability and landslides in the farthest point of the peninsula finding that they are due to the construction of an access road. The impacts registered and the measures taken by a net of wood stacks to measure slope deforestations by wind, rain and continuous car vibrations show an approximate 5% loss in slope vegetation. This research allows the decision makers to implement a series of recommendations into the regulations they have regarding tourist areas to develop as well as access roads.
Keywords: slope instability, landslides, anthropogenic impact.

1 Introduction

The city of Ensenada in Baja California, Mexico is visited annually by approximately 350,000 tourists that either come aboard cruise ships or travel by car. One of its most important tourist attractions is "La Bufadora" ("The Blowhole") which is a famous spurt of sea water (Figure 1) located at the end of the Punta Banda Peninsula.

Figure 1:   La Bufadora (The Blowhole) water spurt.

The road to La Bufadora starts approximately 27 kilometres south of the city of Ensenada towards the west, covering all the arm of the Punta Banda Peninsula facing the Todos Santos Bay.

Punta Banda is about 8 to 9 km in length and 2 to 6 km wide and it is filled with restaurants and small Mexican food stands as well as housing developments for retired Americans, tourists and locals. The road to Punta Banda diverges into several smaller tourists' sites mainly for activities such as fishing, scuba diving and kayaking.

About 5 kilometers before arriving to La Bufadora, there is an exit leading to a new asphalted road towards La Lobera, a major housing resort and golf course development planned for tourists.

1.1 Tectonics

Ensenada is located between the Pacific and North American plate boundary, on the San Andreas Fault zone (Soares and Acosta Chang [1]). One of the faults of this system, the Agua Blanca Fault, a major right-handed strike-slip fault is the main structural element affecting the Punta Banda Peninsula; it is divided in two segments limiting the northern and southern part of the Peninsula.

In 2000, González-Fernández et al. [2] proposed a complex geological model for Punta Banda after interpreting gravity and magnetic data using three dimensional inversion, suggesting the presence of three smaller ruptures along the Punta Banda Peninsula.

Figure 2 shows two of those ruptures located in the study area.

Scarce seismic activity in this region has been registered by the Centro de Investigación Científica y de Educación Superior de Ensenada (CICESE) since 1980, with small swarms of microseismicity occurring approximately every six

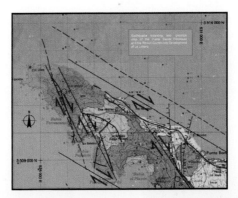

Figure 2: Proposed faults (dashed lines) and interpreted faults (solid lines) for the study area (modified from González-Fernández et al. [2]).

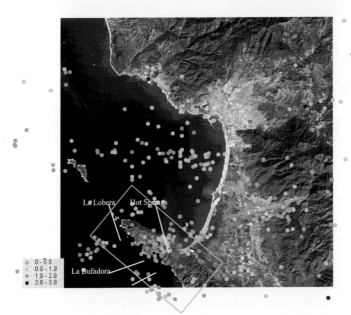

Figure 3: Microseismicity registered from 1982 to 2007 (CICESE [3]). The area in the rectangle comprises the Peninsula of Punta Banda.

months in one section of the main road to La Bufadora, near a hot spring area (Figure 3) and at El Playón Bay which shows evidence of normal faulting.

2 Slope analysis

Punta Banda has a maximum height of approximately 360 meters, sloping gradually to the bay in its north side and having extremely steep cliffs on its south side. Field Surveys were made in August of 2007 to contrast topographic

128 Safety & Security Engineering

maps, gather information of possible rupture zones and to deploy 35 stack poles on La Lobera Resort Development and on its access road to measure possible displacements. Preliminary information from the stacks was collected in January and February of 2008 (Figure 4) and is still being collected.

(a)　　　　　　　　　　　　(b)

Figure 4:　　Stack pole deployment along the La Lobera Development (a) and on the access road to La Lobera Development (b).

A reconnaissance flight was made on February 2008 to have a panoramic view of the study area searching for evidence of fault traces or recent landslides. Aerial photographs were taken along the predetermined course to cover the main areas of interest. In order to track the aircraft position, a GPS unit was used synchronized with the camera to provide an accurate location for each photograph.

2.1 Mudslides and landslides

For three consecutive years, high temperature Santa Ana Winds that arise when high pressure builds over the high plateau regions of Nevada and Utah in the United States of America, blowing towards the Pacific Ocean, produced wildfires in the study area destroying the natural vegetation, with the hillsides remaining vulnerable to weathering and sliding, causing a high risk to the families living in the Punta Banda Peninsula mainly on the Esteban Cantú communal land. Santa Ana winds may blow at approximately at 65 km/h with gusts of 110 km/h with temperatures that can rise over 37^0 C.

During the winter of 2007, runoff water from the storms reached high velocities due to the steep and burned bare slopes lacking the vegetation to hold the soil in place, resulting in flooding and mudslides in the lower areas (Figure 5)

Evidence of anthropogenic impact was found where the addition of loads due to the presence of a tourist housing development on the edges of the hillsides were built in areas where loess and sand forming hills are present (Figure 6), causing additional sliding. In this sector, 4m. of the road to La Bufadora have sunk approximately 5 cm in two years and earthquake swarms have been registered (CICESE [3]).

Figure 5:   Flooding at the Esteban Cantú communal land in Punta Banda.

Figure 6:   Tourist Housing near the area of El Rincón de las Ballenas where normal faulting and erosion due to rain precipitation are present, adding a risk to the residents.

2.2 Roads

In general, the road to La Bufadora subsides on several sectors aligned with the runoff water marks on the hills, even where preventive measures were taken inserting drainpipes in different locations. At the farthest point of the Punta Banda Peninsula, in 2004, an access asphalt road was built for La Lobera Development, where we found that the loads from the construction traffic when acting on the asphalt, generated the immediate responses shown in figure 7 as cracks and sinking and sliding of the road.

A rapid visual screening of the runoff water drainage ways indicates it ends immediately at the side of the road (Figure 7c), which will result in local erosion and could lead to waterlogged roads by mud blocking the pipe when exceptional rainfall delivers great volumes of water. Two of the pipes found were already blocked. It was also found that the cracking of the road is due to the lack of a good soil compaction and the sliding is the result of the removal of the natural vegetation cover by dropping debris material onto the side cross section (Figure 8) destabilizing the hill.

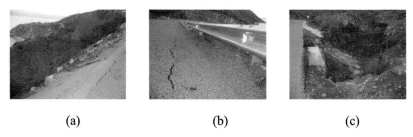

(a) (b) (c)

Figure 7:   Sliding (a), cracking (b) and sinking (c) of the asphalt road to La Lobera.

Figure 8:   Aerial photograph showing debris material dropped to the hillside.

2.3 The model

Based on field recognition and maps, we have classified the slopes as shown in Figure 9 and their analysis is given in Table 1.

In addition, it is strongly recommended as a preventive measure to avoid any building or development at least at a minimum distance of 50 m. to the superficial trace of any fault.

The asphalt road to La Lobera does not fulfill its primary function, causing lack of safety for the users, as well as damage to their vehicles. In time, this could lead to a major rupture or sliding of the pavement if heavy machinery for the tourist development is transiting, leaving the community without any access to the main road.

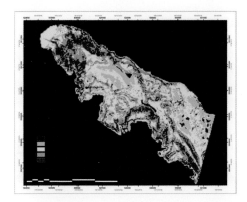

Figure 9: Slope classification in the Punta Banda Peninsula.

Table 1: Slope analysis.

Slope Range (%)	Slope Classification	Land Use Suitability
0 - 0.1	Flat	Suitable for housing. Located on the higher areas. No sliding hazard.
0.1 - 10	Gradual	If located on the higher areas, planning for urban development may be suitable. The lower areas with these slopes are located where loess and sand form the soil and are not recommended for urban development. They are recommended for green areas.
10 - 30	Moderate	Housing development on these slopes could be suitable but costly. They will increase the surface runoff. Not recommended for development where loess and sand forming hills are present.
30 - 60	Steep	Not recommended for housing
60 – 100	Very Steep	Not recommended for housing

In a natural way, the slopes in Punta Banda have been stabilized by native species of vegetation that are resistant to the prolonged droughts and require less water, capturing it mainly from the breeze. The destabilization of the hillsides caused by the removal of the vegetation in the construction of the access road to La Lobera represents a potential hazard specially in the rainy season where water will drain from the upper hill in large quantities and will have more energy, bringing down weathered material as avalanches of mud and the cracking of the road.

3 Proposed interventions and conclusions

The construction of an energy dissipation device on pipes could repair the problem caused by the inadequate drainage system already built; however a constant, preventive and permanent maintenance of the road must be considered due to the inappropriate ground compaction and to the possibility of mudslides from the hills.

To achieve the development of the infrastructure activities on La Lobera it is important to consider the immediate reforestation of the landscape and to protect the remaining one, as a mean to solve problem areas which have been already modified, to protect the hillsides and avoid natural hazards mainly in the rainy season.

It is also important as a safety measure to consider the construction of an alternate route to use in case of an eventuality (v. gr. earthquakes, wildfires) as well as a contingency plan for wildfires and in case it occurs, to promote the fast reforestation for the slope stabilization.

It is highly recommended to build a protected area and a breakwater at La Lobera to secure boats to use in case of any emergencies or illness of the residents.

Slope instability in the study region is mostly due to the pressures caused by the housing developments for tourists. Construction builders should build in areas not susceptible to landslides as recommended in this paper, to ensure the future of the tourism as one of the main variables in the economic growth for Ensenada.

References

[1] Soares J. and J. Acosta Chang. An application of the seismic microzonation to the critical facilities in the city of Ensenada, Baja California, Mexico. Earthquake Resistant Engineering Structures V. WIT Press pp. 63-72. 2005.
[2] A. González-Fernández, B. Martín-Atienza y S. Paz-López. Identificación de fallamiento en la Península de Punta Banda, B. C. a partir de datos de topografía, magnetometría y gravimetría. GEOS. Vol. 20 No 2, pp 98-106. 2000.
[3] Centro de Investigación Científica y de Educación Superior de Ensenada (CICESE). RESNOM Online catalog. http://sismologia.cicese.mx/resnom/catalogo/datain.php

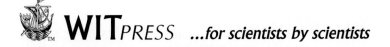

Flood Risk Assessment & Management

Edited by: **S. MAMBRETTI**, *Politecnico Di Milano, Italy*

This volume is the first in a new series that covers various aspects of Safety and Security Engineering with the aim of developing a comprehensive view on risk mitigation. This volume is devoted to floods, since one-third of annual natural disasters and economic losses, and more than half of the victims of natural disasters are flood-related.

The risk from flooding, and the demand for protection from it, has been growing exponentially as a result of a burgeoning global population and growing wealth, climate change and urban development. These factors make it imperative that we change the way flood risk is managed.

Knowledge and scientific tools play a role of paramount importance in the strain of coping with flooding problems, along with capacity building in the context of political and administrative frameworks. Therefore, governments need to establish clear institutional, financial and social mechanisms and processes for flood risk management in order to ensure the safety of people and property and, thereby, contribute to both flood defence and sustainable development.

The present volume contains selected papers presented at Conferences organised by the Wessex Institute of Technology. The papers have been revised by the Authors to bring them up to date and to integrate them into a coherent understanding of the topic. It covers: Risk Assessment; Mathematical Models for Flood Propagation; Effect of Topographic Data Resolution; Social and Psychological Aspects; Decision Making and Management; Legislations and Directives; Alternatives in Flood Protection; Response and Recovery; Damages and Economic-related Problems; Case Studies

The quality of the material makes the volume a most valuable and up-to-date tool for professionals, scientists, and managers to appreciate the state of the art in this important field of knowledge.

Series: Safety & Security Engineering
ISBN: 978-1-84564-646-2   eISBN: 978-1-84564-647-9
Forthcoming 2012 / 144pp / apx £59.00

 WIT*PRESS* ...*for scientists by scientists*

Disaster Management and the Human Health Risk II
Reducing Risk, Improving Outcomes

Edited by: *C.A. BREBBIA*, Wessex Institute of Technology, UK; *A.J. KASSAB*, University of Central Florida, USA and *E.A. DIVO*, Daytona State College, USA

This book contains papers presented at the second International Conference on Disaster Management, held in 2011 in Orlando, Florida. Florida is no stranger to the increasing number of natural disasters affecting millions of people around the world, periodically suffering the ravages of hurricanes, with a recent disastrous sequence of four hurricanes having delivered devastating blows to the State from both Gulf and Atlantic coasts in 2004. Most recently, Florida's economy was seriously affected by the Deep Horizon oil spill in the Gulf of Mexico in spite of its coastline having been mostly spared from any major damage.

The 2011 Disaster Management Conference attracted outstanding contributions from researchers throughout the world. The collected papers, published in this book, reflect the excellent work of all contributing authors and the care taken by the Scientific Advisory Committee and other colleagues in reviewing the presentations. Topics covered include: Disaster Analysis, Monitoring and Mitigation; Emergency Preparedness; Risk Mitigation; Surveillance and Early Warning Systems; Socio-Economic Issues.

WIT Transactions on The Built Environment, Vol 119
ISBN: 978-1-84564-536-6   eISBN: 978-1-84564-537-3
Published 2011 / 336pp / £145.00

All prices correct at time of going to press but subject to change.
WIT Press books are available through your bookseller or direct from the publisher.

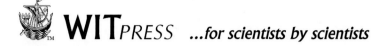
...for scientists by scientists

Monitoring, Simulation, Prevention and Remediation of Dense and Debris Flows IV

Edited by: **D. DE WRACHIEN**, Politecnico di Milano, Italy; **C.A. BREBBIA**, Wessex Institute of Technology, UK and **S. MAMBRETTI**, Politecnico di Milano, Italy

Containing papers presented at the fourth and latest in a series of biennial International Conferences dealing with the Monitoring, Simulation, Prevention and Remediation of Dense and Debris Flow, this book addresses the need for a better understanding of the ever more frequent phenomena known as debris flows.

Due to the increased occurrence of debris and hyper-concentrated flows and the impact they have on both the environment and human life, these extreme events and the related processes have been attracting increased attention from research groups and land planning and management professionals. A full understanding of these phenomena will lead to an integrated risk management approach that can include measures to prevent a hazard turning into a disaster. The papers presented at the conference deal with: Debris Flow Modelling; Debris Flow Triggering; Risk Assessment; Hazard Mitigation; Sediment Transport and Debris Flow Monitoring and Analysis; Landslide Phenomena; Debris Flow Rheology; Active and Passive Disaster Management; Vulnerability Studies; Structural and Non-structural Controls; Field Tests. The book will be useful to engineers, scientists and managers from laboratories, industry, government and academia who deal with risk management, natural disasters, and the phenomenon itself.

WIT Transactions on Engineering Sciences, Vol 73
ISBN: 978-1-84564-586-1   eISBN: 978-1-84564-587-8
Forthcoming 2012   /   apx 300pp   /   apx £129.00

WITPress
Ashurst Lodge, Ashurst, Southampton,
SO40 7AA, UK.
Tel: 44 (0) 238 029 3223
Fax: 44 (0) 238 029 2853
E-Mail: witpress@witpress.com

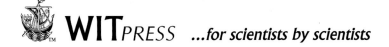

Safety and Security Engineering IV

Edited by: **M. GUARASCIO**, *University of Rome 'La Sapienza', Italy;* **G. RENIERS**, *University of Antwerp, Belgium;* **C.A. BREBBIA**, *Wessex Institute of Technology, UK and* **F. GARZIA**, *University of Rome 'La Sapienza', Italy*

Safety and Security Engineering, due to its special nature, represents an interdisciplinary area of research and applications that brings together, in a systemic view, many disciplines of engineering, from the most traditional to the most advanced and novel.

Safety and Security Engineering is characterised by a totally new approach since it first analyses the hazard context, not only by means of traditional tools but also by means of risk analysis techniques, and then manages the above-mentioned context through technical solutions, installations, systems, human resources and procedures to prevent and face incidental events, natural and voluntary, that could damage people or goods.

The Fourth International Conference on Safety and Security Engineering (SAFE 2011) was convened to present and discuss the most recent developments in the theoretical and practical aspects of Safety and Security Engineering. The Conference papers in this volume cover the following topics: Infrastructure Protection; Risk Analysis, Assessment and Management; Public Safety and Security; Modelling and Experiments; Construction Safety and Security; Transportation and Road Safety; Safety of Users in Road Evacuation; Emergency and Disaster Management; Process Safety and Security; Emerging Issues in Safety.

WIT Transactions on The Built Environment, Vol 117
ISBN: 978-1-84564-522-9   eISBN: 978-1-84564-523-6
Published 2011 / 544pp / £234.00